Wavelets and Multi-Resolution Analysis

M S Sinith • Gayathri A • Chithra K R

Wavelets and Multi-Resolution Analysis

A Signal Processing Approach

Springer

M S Sinith
Dept. of Electronics and Communication Engineering
Government Engineering College Wayanad
Wayanad, Kerala, India

Gayathri A
Dept. of Electronics and Communication Engineering
Government Engineering College Palakkad
Palakkad, Kerala, India

Chithra K R
Dept. of Electronics and Communication Engineering
Government Engineering College Thrissur
Thrissur, Kerala, India

ISBN 978-3-031-92749-2 ISBN 978-3-031-92750-8 (eBook)
https://doi.org/10.1007/978-3-031-92750-8

This Springer imprint is published by the registered company Springer Nature Switzerland AG
The registered company address is: Gewerbestrasse 11, 6330 Cham, Switzerland

To my daughter Chinmayi, my Son Achyuth and all other children, who are the promise of tomorrow

Sinith

To my parents,
for your love and sacrifices.
To my teachers,
for your guidance and wisdom
To my husband,
for your constant support and belief in me.

Gayathri

To my parents and teachers, the guiding lights on my path
To my husband and children, the greatest blessings of my life

Chithra

Contents

Chapter 1
Signal Space Concept

1.1 Vector Space

A vector space is a space where all the dimensions are vectors. The linear combinations of any vectors give a new vector which is part of the vector space. Vector space can be defined as the space obtained by including all the vectors generated by all possible linear combinations of the unit vectors of the space. The dimensionality of the vector space can vary from unity to infinity.

One Dimensional Vector Space

Any point in a one dimensional vector space can be represented in terms of the unit vector in that direction as

$$\mathbf{A} = a\hat{i} \tag{1.1}$$

where a is a scalar and $\hat{i}$ is the unit vector in that direction (Fig. 1.1).

Two Dimensional Vector Space

Any point $\mathbf{A}$ in a two dimensional space can be represented as

$$\mathbf{A} = c\hat{i} + d\hat{j} \tag{1.2}$$

where c and d are scalars and $\hat{i}$ and $\hat{j}$ are unit vectors in two dimensions. The resultant vector usually can be represented in a plane (Fig. 1.2).

M. S. Sinith et al., *Wavelets and Multi-Resolution Analysis*,
https://doi.org/10.1007/978-3-031-92750-8_1

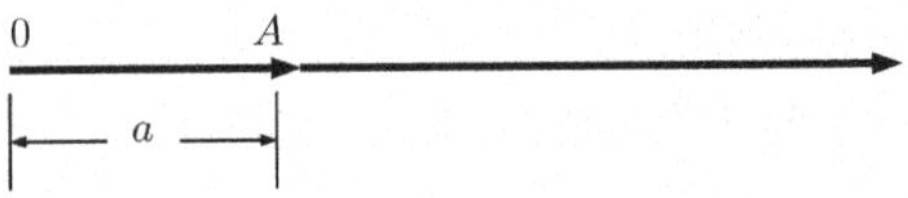

Fig. 1.1 Representation of a point in one dimensional vector space

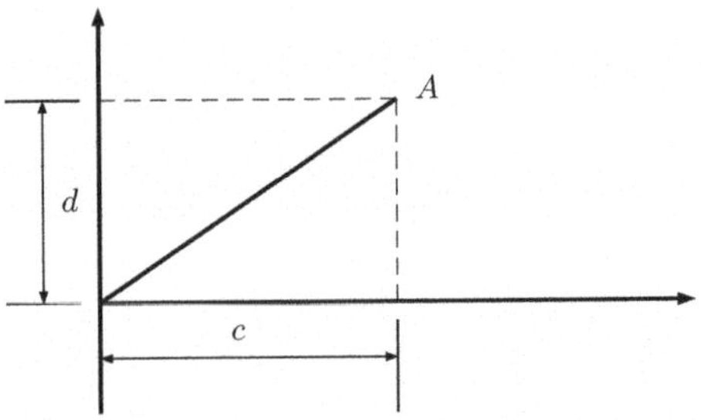

Fig. 1.2 Representation of a point in two dimensional vector space in a plane

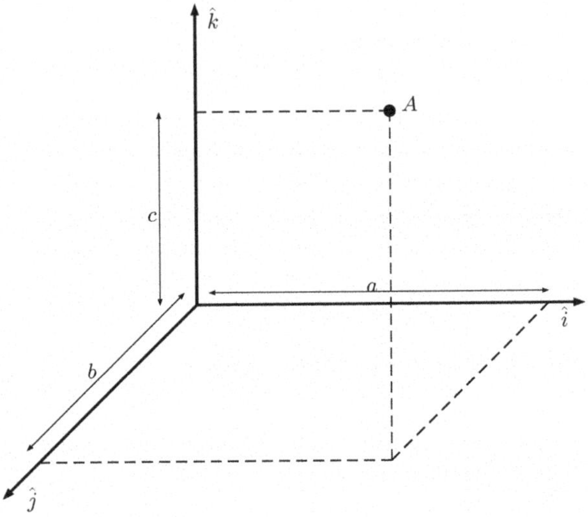

Fig. 1.3 Representation of a point in three-dimensional vector space

Three-Dimensional Vector Space

Any point **A** in a three-dimensional space can be represented as

$$\mathbf{A} = a\hat{i} + b\hat{j} + c\hat{k} \tag{1.3}$$

where a, b, and c are scalars and are also called the projections of the vector **A** to x-axis, y-axis, and z-axis, respectively, and $\hat{i}$, $\hat{j}$, and $\hat{k}$ are the unit vectors in the direction of x-axis, y-axis, and z-axis, respectively (Fig. 1.3).

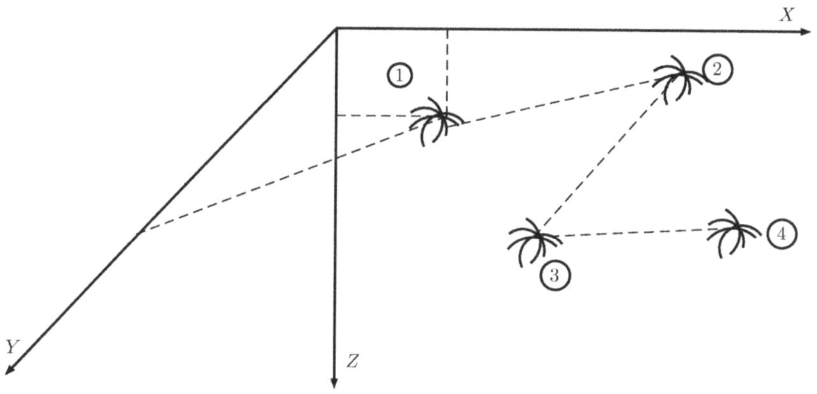

Fig. 1.4 Example of a three-dimensional space

N-Dimensional Vector Space

Let us consider an abandoned room in which a spider moves through its web with a constant velocity making its web. If we consider the position of the spider at any time t, it can be represented as $S = ut$. If we represent the positions 1, 2, 3, and 4 as t_1, t_2, t_3, and t_4, then the time at which the spider is at:

the position at time t_1 is $a_1\hat{i} + b_1\hat{j} - c_1\hat{k}$
the position at time t_2 is $a_2\hat{i} + b_2\hat{j} - c_2\hat{k}$
the position at time t_3 is $a_3\hat{i} + b_3\hat{j} - c_3\hat{k}$

where a_i, b_i, and c_i represent the projection of the positions to the x, y, and z directions at time t_i (Fig. 1.4). Now the four-dimensional representations of these vectors are (a_i, b_i, c_i and ut_i).

In continuous time, this forms a four-dimensional vector space. Similar to this, a vector space can have any dimensions, including infinite dimensions.

1.2 Signal Space

A signal space is a space similar to the vector space, where the vectors are replaced by signals.

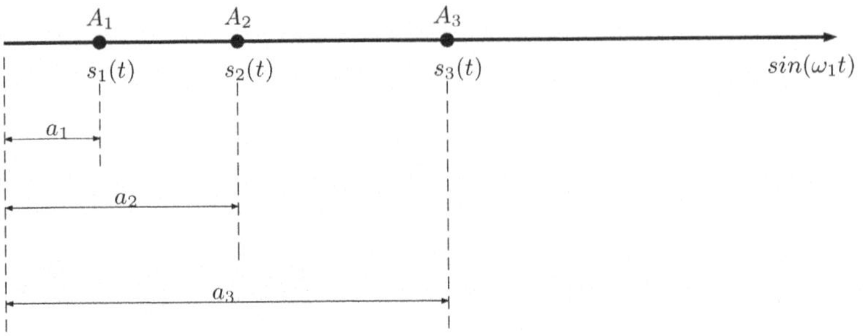

Fig. 1.5 Signals in a single-dimensional vector space

Fig. 1.6 Signals in a two dimensional vector space

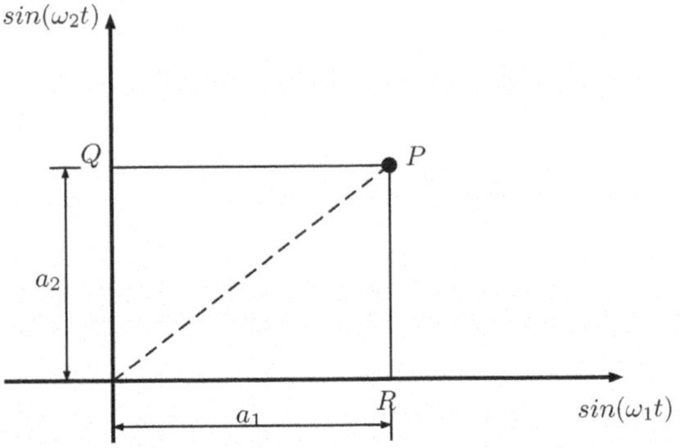

N-Dimensional Signal Space

In a single-dimensional signal space represented in Fig. 1.5, points A_1, A_2 and A_3 represent three different signals:

$$s_1(t) = a_1 \sin(\omega_1 t)$$
$$s_2(t) = a_2 \sin(\omega_1 t)$$
$$s_3(t) = a_3 \sin(\omega_1 t)$$

In precise, the three signals have same frequency, while they have different amplitudes. As in the case of unit vectors in vector space, normalized signals (with unit energy or power) become unit signals. Here $\sin(\omega_1 t)$ should be normalized.

In a two dimensional signal space, the two dimensions shall be two signals satisfying orthogonality conditions. In Fig. 1.6, Q and R represent sine waves of different frequencies and amplitudes where P represents a periodic signal with two spectral peaks at ω_1 and ω_2.

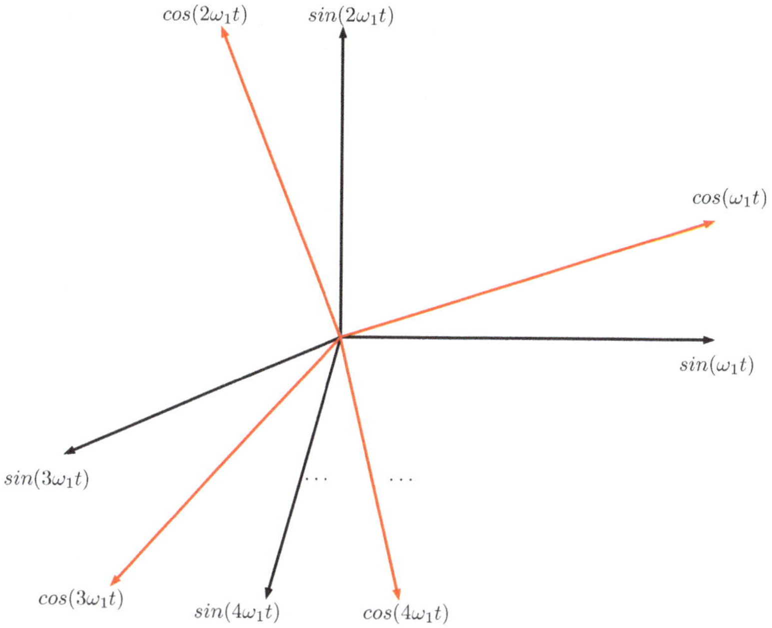

Fig. 1.7 Infinite dimensional signal space

Let us now consider the case of Fourier series. Fourier series can be used to represent any periodic waveform as a sum of sinusoidals of different frequencies and amplitudes. Any periodic signal $f(t)$ can be represented as

$$f(t) = a_0 + \sum_{n=1}^{\infty} a_n \cos(\omega_n t) + \sum_{n=1}^{\infty} \sin(\omega_n t) \tag{1.4}$$

where a_0 is the DC component; a_1, a_2, a_3....... and b_1, b_2, b_3, ... are the amplitudes of different harmonics.

Figure 1.7 shows an infinite dimensional signal space with $\sin(\omega_1 t)$, $\sin(2\omega_1 t)$, $\sin(3\omega_1 t)$... and $\cos(\omega_1 t)$, $\cos(2\omega_1 t)$, $\cos(3\omega_1 t)$.... forming the orthogonal axes. In this case, any point in this infinite dimensional space represents a periodic signal. The projection of the point to each axis gives the spectral component corresponding to the frequency represented in that particular axis.

1.3 Inner Product

Any operation $\langle -, - \rangle$ in a field k can be defined as inner product if all the four properties given below are satisfied:

1. $\langle x, y + z\rangle = \langle x, y\rangle + \langle x, z\rangle$ (*distributive property*).
2. $\langle x, \lambda y\rangle = \bar{\lambda}\langle x, y\rangle$ (*conjugate symmetry property*).
3. $\langle x, 0\rangle = \langle 0, x\rangle = 0$.
4. If $\langle x, z\rangle = 0$ for all values of z, then $x = 0$.

Inner Product in Vector Space

An inner product that satisfies the properties mentioned in Sect. 1.3 is dot product. Dot product between two vectors is defined as

$$\langle \mathbf{x}, \mathbf{y}\rangle = \langle \mathbf{x}.\mathbf{y}\rangle =\mid \mathbf{x} \mid\mid \mathbf{y} \mid \cos\theta \tag{1.5}$$

where $\mid \mathbf{x} \mid$ and $\mid \mathbf{y} \mid$ are norms and θ is the angle between the vectors **x** and **y**.

Inner Product in Signal Space

An operation that satisfies the properties of inner product is

$$\int_0^T x(t)y(t)dt$$

where $x(t)$ and $y(t)$ are signals. If the inner product between two signals is zero, they are said to be orthogonal to each other,

$$\text{i.e.,}\quad \text{if} \int_0^T x(t)y(t)dt = 0 \tag{1.6}$$

then $x(t)$ is said to be orthogonal to $y(t)$. It can be easily shown that

$$\int_0^T \sin\left(\frac{2\pi}{T}t\right)\cos\left(\frac{2\pi}{T}t\right)dt = 0 \tag{1.7}$$

1.4 Orthonormal Basis Functions

Basis Functions

A basis function or signal is defined as the set of functions or signals which can span a given space. In other words, any arbitrary signal in the space should be able to be represented using a linear combination of the basis.

Orthonormal Signals

Any two signals are called orthonormal, if they are orthogonal to each other and normal (norm = 1). For example, $x(t)$ and $y(t)$ form orthonormal basis functions if all the properties given below are satisfied.

$$\begin{aligned} \int_0^T x(t)y(t)dt &= 0 \\ \int_0^T \mid x^2(t) \mid dt &= 1 \\ \int_0^T \mid y^2(t) \mid dt &= 1 \end{aligned} \tag{1.8}$$

In addition, if their linear combinations span the given signal space, they are called orthonormal basis functions.

1.5 Solved Problems

Problem Verify if the set of all vectors of the form $a\hat{i} + b\hat{j}$, where a and b are real numbers, forms a vector space.

Solution We verify the vector space axioms:

- Closure under addition and scalar multiplication: Holds.
- Commutativity and associativity: Hold for addition of real numbers.
- Existence of zero vector and additive inverses: Holds.
- Distributivity and compatibility: Hold.

Thus, the set forms a vector space.

Problem If $\mathbf{A} = a\hat{i} + b\hat{j} + c\hat{k}$, find the scalar projections onto the axes.

Solution

- Scalar projection onto the x-axis: a
- Scalar projection onto the y-axis: b
- Scalar projection onto the z-axis: c

Problem Consider a spider's position in 3D space at times t_1 and t_2. Represent as a 4D vector and calculate the vector sum.

Solution Let the positions be $\mathbf{P_1} = \begin{pmatrix} x_1 \\ y_1 \\ z_1 \\ t_1 \end{pmatrix}$ and $\mathbf{P_2} = \begin{pmatrix} x_2 \\ y_2 \\ z_2 \\ t_2 \end{pmatrix}$. The vector sum is

$$\mathbf{P_1} + \mathbf{P_2} = \begin{pmatrix} x_1 + x_2 \\ y_1 + y_2 \\ z_1 + z_2 \\ t_1 + t_2 \end{pmatrix}$$

Problem Consider the signal $f(t) = 3 + 2\sin(2t) + 4\cos(3t)$. Identify spectral components.

Solution Spectral components:

- DC component: 3
- $\sin(2t)$ component: 2
- $\cos(3t)$ component: 4

Problem Compute the inner product $\langle x, y \rangle$ for $x(t) = 2\sin(2\pi t)$ and $y(t) = 3\cos(2\pi t)$.

Solution

$$\langle x, y \rangle = \int_0^1 2\sin(2\pi t) \cdot 3\cos(2\pi t)\, dt = 0$$

Since the inner product is zero, $x(t)$ and $y(t)$ are orthogonal.

Problem Verify the properties of inner product for given three vectors in $\mathbb{R}^3$

$$\mathbf{u} = \begin{bmatrix} 1 \\ 2 \\ 3 \end{bmatrix}, \quad \mathbf{v} = \begin{bmatrix} 4 \\ 5 \\ 6 \end{bmatrix}, \quad \mathbf{w} = \begin{bmatrix} 7 \\ 8 \\ 9 \end{bmatrix}$$

Solution

1. **Commutativity**
   The inner product is given by

$$\langle \mathbf{u}, \mathbf{v} \rangle = 1(4) + 2(5) + 3(6) = 4 + 10 + 18 = 32$$

   Similarly,

$$\langle \mathbf{v}, \mathbf{u} \rangle = 4(1) + 5(2) + 6(3) = 4 + 10 + 18 = 32$$

   Since $\langle \mathbf{u}, \mathbf{v} \rangle = \langle \mathbf{v}, \mathbf{u} \rangle$, commutativity holds.

2. **Distributivity**
   First, compute $\mathbf{u} + \mathbf{v}$:

$$\mathbf{u} + \mathbf{v} = \begin{bmatrix} 1+4 \\ 2+5 \\ 3+6 \end{bmatrix} = \begin{bmatrix} 5 \\ 7 \\ 9 \end{bmatrix}$$

   Now compute

$$\langle \mathbf{u} + \mathbf{v}, \mathbf{w} \rangle = 5(7) + 7(8) + 9(9) = 35 + 56 + 81 = 172$$

   Separately,

$$\langle \mathbf{u}, \mathbf{w} \rangle = 1(7) + 2(8) + 3(9) = 7 + 16 + 27 = 50$$

$$\langle \mathbf{v}, \mathbf{w} \rangle = 4(7) + 5(8) + 6(9) = 28 + 40 + 54 = 122$$

   Since $\langle \mathbf{u} + \mathbf{v}, \mathbf{w} \rangle = 50 + 122 = 172$, distributivity holds.
3. **Scalar Multiplication**
   Let $c = 2$, and then:

$$\langle 2\mathbf{u}, \mathbf{v} \rangle = \langle \begin{bmatrix} 2(1) \\ 2(2) \\ 2(3) \end{bmatrix}, \mathbf{v} \rangle = \langle \begin{bmatrix} 2 \\ 4 \\ 6 \end{bmatrix}, \mathbf{v} \rangle$$

$$= 2(4) + 4(5) + 6(6) = 8 + 20 + 36 = 64$$

   Also,

$$2\langle \mathbf{u}, \mathbf{v} \rangle = 2 \times 32 = 64$$

   Since $\langle 2\mathbf{u}, \mathbf{v} \rangle = 2\langle \mathbf{u}, \mathbf{v} \rangle$, scalar multiplication holds.

Problem For $x(t) = 2\sin(2\pi t)$ and $y(t) = 3\sin(4\pi t)$:

(a) Find the projection of $x(t)$ onto $y(t)$.
(b) Significance of the projection.

Solution

(a) The projection of $x(t)$ onto $y(t)$.

$$\langle x(t), y(t) \rangle = 0, \quad \langle y(t), y(t) \rangle = \frac{9}{2}$$

$$\text{Proj}_y(x) = 0$$

(b) Significance of the projection: Since the projection is zero, $x(t)$ and $y(t)$ are orthogonal, which means they represent distinct frequency components.

Chapter 2
Introduction to Wavelet Transform

2.1 Fourier Transform

Consider a crowd which contains people speaking different languages. Suppose that a person addresses the crowd in the following way:

- Bhai behanaum, namaskaar.
- Ladies and Gentlemen, Hai to all.
- Neu shen mein shen shen mein thaja hau.
- Mesdames et messieurs, salut à tous.

Those who understand Hindi will respond to the first sentence. Those who understand English will respond to the second sentence, those who understand Chinese will respond to the third sentence, and those who understand French will respond to the fourth sentence.

Now, consider the equation for finding Fourier transform,

$$X(j\omega) = \int_{-\infty}^{\infty} x(t)e^{-j\omega t}dt \tag{2.1}$$

Similar to the situation mentioned just above, just as the people respond to a particular language they understand, Fourier transform, X($j\omega$) also responds to a

M. S. Sinith et al., *Wavelets and Multi-Resolution Analysis*,
https://doi.org/10.1007/978-3-031-92750-8_2

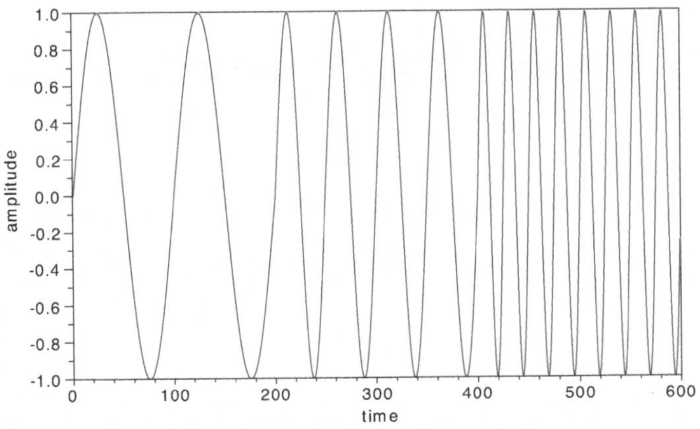

Fig. 2.1 A sample signal

particular frequency ω_0 that is present in the signal x(t). Mathematically,

$$X(j\omega_0) = \int_{-\infty}^{\infty} x(t)e^{-j\omega_0 t}dt \tag{2.2}$$

Thus if ω takes all values from $-\infty$ to ∞, we get Eq. 2.1. In Eq. 2.1, the basis function used is $e^{-j\omega t}$, which is basically a sinusoid (real part). Hence, Eq. 2.1 is the correlation between the signal $x(t)$ and sinusoidals whose frequency varies continuously from $-\infty$ to ∞. Now consider the signal shown in Fig. 2.1, which contains three sinusoidals having frequencies f_1, f_2, f_3 and which exists from 0 to 200 units of time, from 200 to 400 units of time, and from 400 to 600 units of time, respectively. The Fourier transform of this signal, found using Eq. 2.1, will have three broadened impulse functions (since the sinusoidals are truncated in time domain) at f_1, f_2, f_3, respectively, as shown in Fig. 2.2. Is there any other signal which has the same Fourier transform as that of the signal shown in Fig. 2.1? The answer is yes. The signal shown in Fig. 2.3 is a signal obtained by adding the three sinusoidals having frequencies f_1, f_2, f_3. The difference between Figs. 2.1 and 2.3 is that, in Fig. 2.1, different sinusoidals exist at different duration, while the signal in Fig. 2.3 has all the frequencies at all time under consideration.

From the discussions, it is clear that Fourier transform is not unique. We do not get the slightest idea *"when"* the signal exists. In other words, the time resolution is zero. Then, what is the solution? Short Time Fourier Transform is the solution.

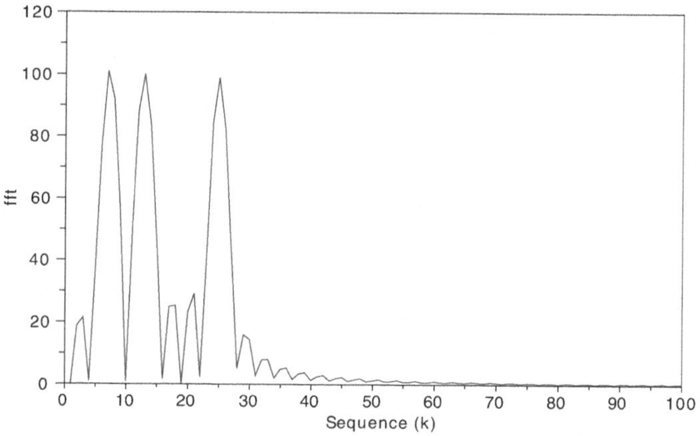

Fig. 2.2 FFT of the signal in Fig. 2.1

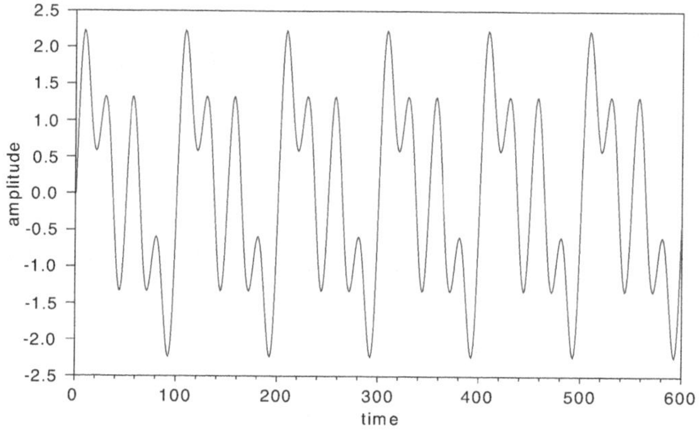

Fig. 2.3 Sum of three sinusoidals

MATLAB Code

```
%...Program to find and plot Fourier transform of the signals shown in Fig.2.1...%
x1=sin(2*3.1415*n);
x2=sin(4*3.1415*n);
x3=sin(8*3.1415*n);
y=[x1 x2 x3];
xlabel('Sequency(k)');
ylabel('fft')
Y=abs(fft(y));
plot(Y(1:100));   % FFT of the signal
% Output of the code is shown in Fig.2.2%
```

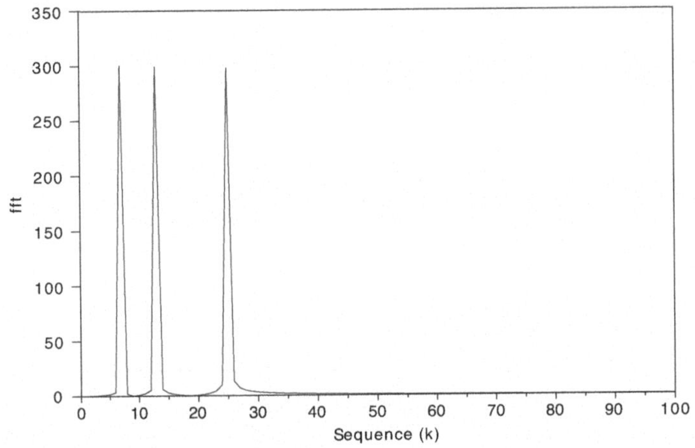

Fig. 2.4 FFT of the signal in Fig. 2.3

MATLAB Code

```
%....Program to find and plot Fourier transform of the signal shown in Fig.2.3.%
n=0:0.01:6;
x1=sin(2*3.1415*n);
x2=sin(4*3.1415*n);
x3=sin(8*3.1415*n);
y=x1+x2+x3;
xlabel('Sequency(k)');
ylabel('fft');
Y=abs(fft(y));
plot(Y(1:100));  % FFT of the signal
% Output of the code is shown in Fig.2.4
```

Note: It should be noted that Figs. 2.2 and 2.4 are a little different because the truncation length for the sinusoidals in Fig. 2.3 is three times as that in Fig. 2.1.

2.2 Short Time Fourier Transform (STFT)

The shortcomings of Fourier transform can be compensated to an extent by taking Fourier transform frame by frame. Mathematically,

$$STFT(t,\tau) = \int_{-\infty}^{\infty} x(t)w(t-\tau)e^{-j\omega t}\,dt \tag{2.3}$$

where $w(t)$ and $w(t-\tau)$ are the window function and its shifted versions. The window function can be a simple rectangular window, Hamming window, Hanning window, or any other suitable window depending on the application. The shifts can be chosen such that consecutive windows have certain overlapping. In some applications, overlapping upto 75% is used. Such overlapping gives a better overall time resolution. The frequency resolution does not get lowered here because the window width is not changed.

Is STFT enough? STFT increased the time resolution in the analysis of signals as compared to Fourier transform. That means, we can identify the time interval at which a particular frequency exists in the signal. Even then we cannot find the exact time. Since the window width is fixed, the time-frequency resolution remains unchanged. We will see in Chapter 3 that, as time resolution increases, frequency resolution decreases and vice versa. To analyze the signal in Fig. 2.1, the optimum window width required for the first few cycles is greater than that is required for the last few cycles. Hence to analysis such signals, we need variable time-frequency resolution. Variable time-frequency resolution can be introduced in the analysis through wavelet transform. The spectrogram of an audio signal obtained from FFT with a Hamming window and with 50% of overlapping in the consecutive windows is shown in Fig. 2.5.

Now consider the signal shown in Fig. 2.6. It is clear from the figure that the best time-frequency resolution will be obtained for the signal till 200 *units* of time, if

Fig. 2.5 STFT of an audio signal

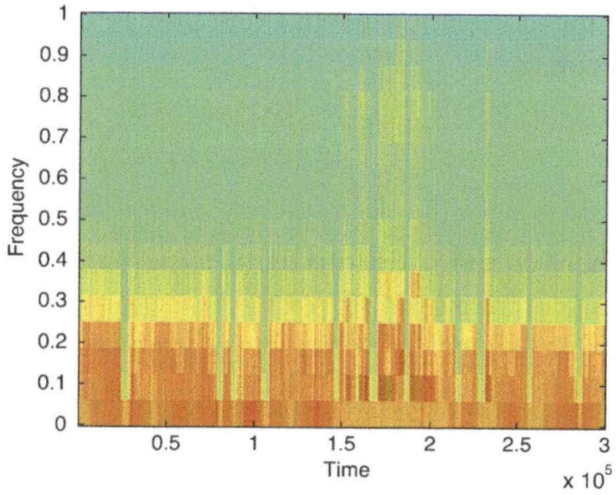

Fig. 2.6 A sample signal

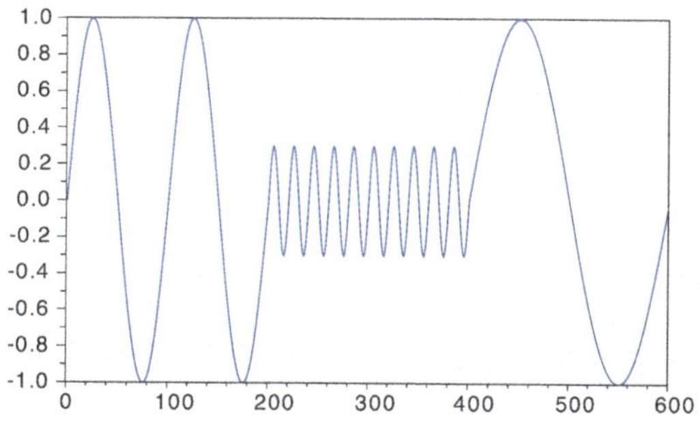

the window width used is of duration 100 *units*. Similarly, the best time-frequency resolution will be obtained for the signal from 200 *units* of time to 400 *units*, if the window width used is of duration 20 *units*. But the window size cannot be varied in STFT. Thus we need a better analyzing tool for obtaining variable time-frequency resolution. This is possible through continuous wavelet transform.

MATLAB Code

```
%....Matlab program for plotting the spectrogram of an audio file....%
x=wavread('song.wav');  % 'song.wav'  should be in the working folder
specgram(x);        % The output of the code is shown in Fig.2.5.
```

2.3 Continuous Wavelet Transform (CWT)

In the previous section, the basis function used is $e^{-j\omega t}$ windowed using $w(t)$, which is basically a truncated sinusoid. If this function is changed with a finite energy signal $\psi(t)$, the equation becomes more generic.

Equation 2.3 now becomes

$$y(t) = \int_{-\infty}^{\infty} x(t)\psi(t)dt \tag{2.4}$$

If the correlation of the function is done in different scales and shifts, Eq. 2.4 becomes

$$CWT(a,b) = \int_{-\infty}^{\infty} x(t)\psi\left(\frac{t-b}{a}\right)dt \tag{2.5}$$

where a and b denote the scaling and shifting factors. This is called continuous wavelet transform (CWT).

The signal $x(t)$ can be reproduced from $CWT(a,b)$ by finding the inverse wavelet transform,

$$x(t) = \int_{-\infty}^{\infty}\int_{-\infty}^{\infty} CWT(a,b)\psi\left(\frac{t-b}{a}\right)da\,db \tag{2.6}$$

Now consider *Haar* wavelet, which is the simplest wavelet (box shaped), defined by the equation

$$\phi(t) = 1 \;\forall\, 0 \leq t \leq 1 \tag{2.7}$$

where $\phi(t)$ is defined as scaling function which is the building block of wavelet function, $\psi(t)$. The scaling function and wavelet function of Haar wavelet are shown in Figs. 2.7a and b.

Fig. 2.7 Haar scaling function and wavelet function

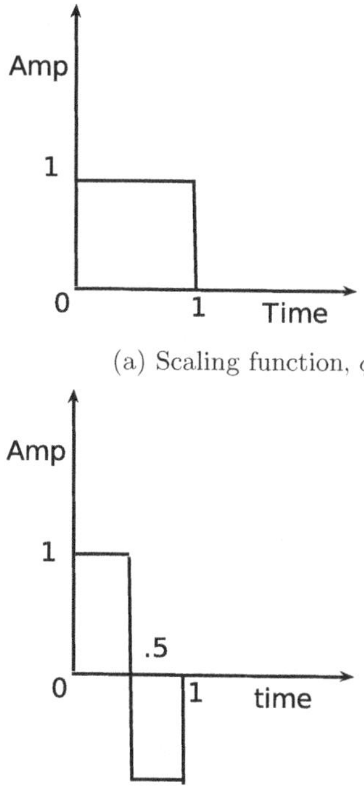

(a) Scaling function, $\phi(t)$

(b) Wavelet function, $\psi(t)$

From Fig. 2.8b, it is clear that $\psi(t)$ can be represented in terms of scaled version of $\phi(t)$ as in Eq. 2.8.

$$\psi(t) = \phi(2t) - \phi(2t - 1) \tag{2.8}$$

In a similar way, as shown in Fig. 2.8a, $\phi(t)$ can also be represented in terms of scaled version of $\phi(t)$. Mathematically,

$$\phi(t) = \phi(2t) + \phi(2t - 1) \tag{2.9}$$

Now, let us make a man using Haar wavelet and its scaled and shifted versions in two dimensions. Figure 2.9a is the basic approximation of a man. When we add some details in the next level, we get Fig. 2.9b. When some more details are added in the consecutive levels, we get Figs. 2.9c and d. The more the number of levels of details, the more perfect the man will become. The scaling function, $\phi(t)$, is used to represent the approximations (low frequencies) in the figure, while the wavelet function, $\psi(t)$, is used to represent the details (high frequencies) in the figure.

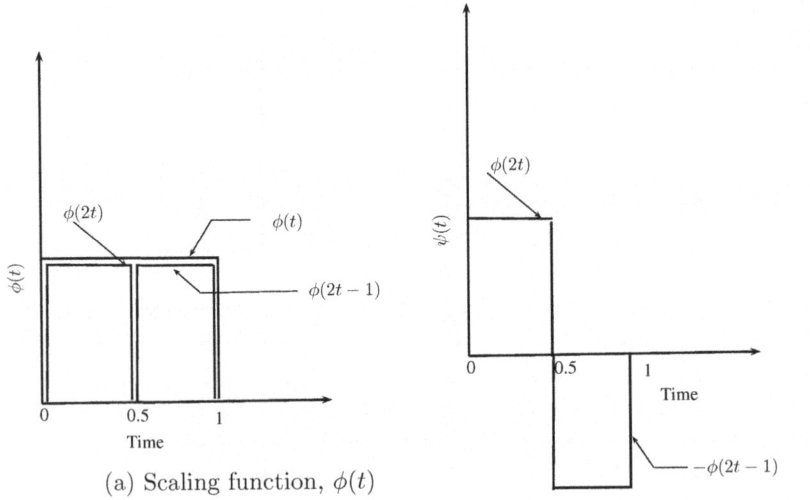

(a) Scaling function, $\phi(t)$

(b) Wavelet function, $\psi(t)$

Fig. 2.8 Haar scaling function and wavelet function

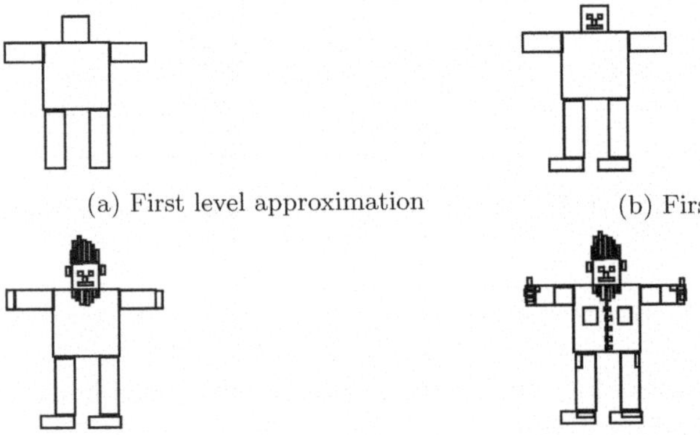

(a) First level approximation

(b) First level detail added

(c) Second level detail added

(d) Third level detail added

Fig. 2.9 A man built using Haar wavelet

Now consider the scaling function, $\phi(t)$ of the *Hat* wavelet, shown in Fig. 2.10. From the figure, it is clear that

$$\phi(t) = \frac{1}{2}\phi(2t) + \phi(2t-1) + \frac{1}{2}\phi(2t-2) \tag{2.10}$$

This equation can be rewritten as

$$\phi(t) = h(0)\phi(2t) + h(1)\phi(2t-1) + h(2)\phi(2t-2) \tag{2.11}$$

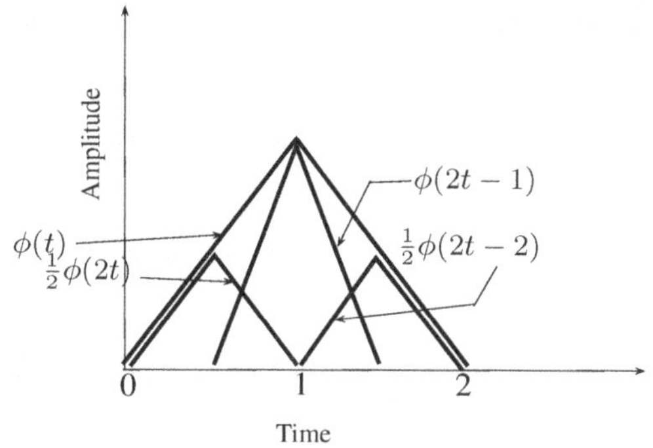

Fig. 2.10 Hat scaling function

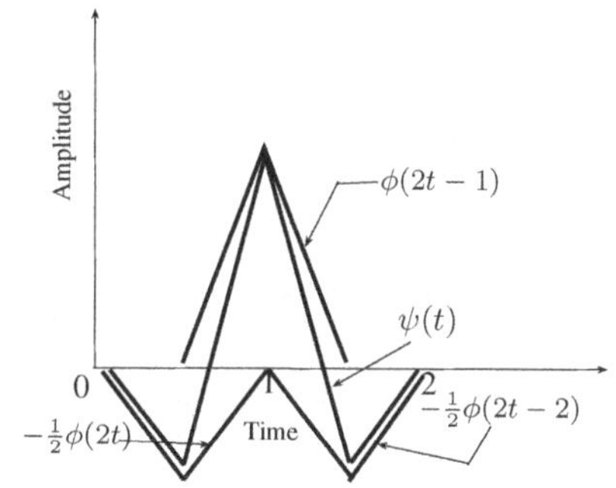

Fig. 2.11 Hat wavelet function

where $h(0) = \frac{1}{2}$, $h(1) = 1$, and $h(2) = \frac{1}{2}$. Now consider the wavelet function, $\psi(t)$ of the *Hat* wavelet, shown in Fig. 2.11. From the figure, it is clear that

$$\psi(t) = -\frac{1}{2}\phi(2t) + \phi(2t-1) + -\frac{1}{2}\phi(2t-2) \tag{2.12}$$

This equation can be rewritten as

$$\psi(t) = g(0)\phi(2t) + g(1)\phi(2t-1) + g(2)\phi(2t-2) \tag{2.13}$$

where $g(0) = -\frac{1}{2}$, $g(1) = 1$, and $g(2) = -\frac{1}{2}$.

2.4 Discrete Wavelet Transform (DWT)

Is CWT enough? Can CWT be implemented in computers and other digital devices? No, the Continuous Wavelet Transform cannot be implemented in computers as there could be infinite shifts even for small amplitude signals and finite energy signals. Now the solution is to have discrete shifts and discrete scales for wavelet which in turn produce a family of daughter wavelets. In discrete wavelet transform (DWT), the shifts and scales are discrete. The scales are diadic (multiples of two)

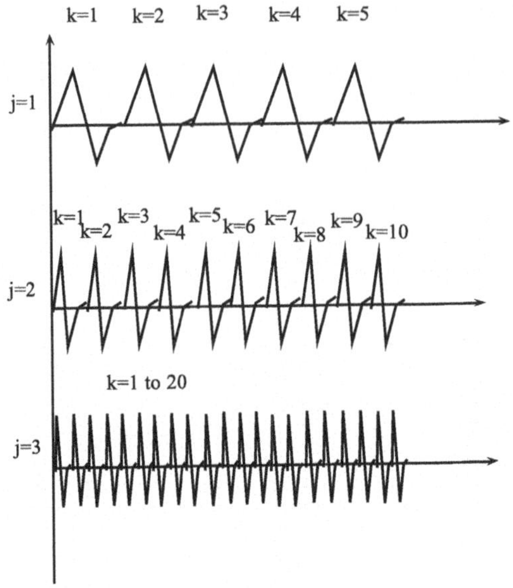

Fig. 2.12 Diadic representation of wavelets

too. Mathematically, the daughter wavelets can be represented as

$$\psi_{j,k}(t) = 2^{\frac{j}{2}} \psi(2^j t - k) \; \forall \; j, k \in Z \tag{2.14}$$

The term $2^{\frac{j}{2}}$ is to normalize the energy of the function. In short, if $\psi(t)$ has unit energy, then all the daughter wavelets also have unit energy.

A sample wavelet and its daughter wavelets for different shifts and scales are shown in Fig. 2.12. It can be easily verified from the figure that the shifted versions of the daughter wavelets at a given scale are orthogonal to each other. Any arbitrary causal function f(t) can be representation in terms of $\phi(t)$ (approximation) and $\psi(t)$ (details) as

$$f(t) = \sum_{k=0}^{\infty} c_k \phi(t-k) + \sum_{j=0}^{\infty} \sum_{k=0}^{\infty} 2^{\frac{j}{2}} d_{j,k} \psi(2^j t - k) \tag{2.15}$$

where c_k and $d_{j,k}$ are the discrete wavelet transform (DWT) coefficients.

$\phi(t)$ and their shifted versions are orthonormal iff

$$\int_{-\infty}^{\infty} \phi(t-k)\phi(t-l)dt = \delta(k-l) \; \forall \; k, l \in Z \tag{2.16}$$

Further, if $\phi(t)$, $\psi(t)$, $\psi_j(t)$, and their shifted versions are orthogonal to each other, normalized, and have zero values for $t \leq 0$, we get the following equations:

$$\int_0^\infty \psi(t-k)\psi(t-l)dt = \delta(k-l)\ \forall\, k, l \in Z \tag{2.17}$$

$$\int_0^\infty \phi(t-k)\psi(t-l)dt = 0\ \forall\, k, l \in Z \tag{2.18}$$

$$\int_0^\infty \phi(t-k)2^{\frac{j}{2}}\psi(2^j t-l)dt = 0\ \forall\, k, l \in Z \tag{2.19}$$

$$\int_0^\infty 2^{\frac{j}{2}}\psi(2^j t-k)2^{\frac{j}{2}}\psi(2^j t-l)dt = \delta(k-l)\ \forall\, k, l \in Z \tag{2.20}$$

Multiplying Eq. 2.15 with $\phi(t-k)$, integrating w.r.t. t, and using Eqs. 2.16 and 2.19

$$c_k = \int_0^\infty f(t)\phi(t-k)dt \tag{2.21}$$

Similarly, using Eqs. 2.15, 2.19, and 2.20,

$$d_{j,k} = \int_0^\infty f(t)2^{\frac{j}{2}}\psi(2^j t-k)dt \tag{2.22}$$

Thus the DWT coefficient, c_k, of a signal can be obtained by multiplying the signal with scaling function and integrating w.r.t. t. Similarly, the DWT coefficient, $d_{j,k}$, of a signal can be obtained by multiplying the signal with wavelet function and integrating w.r.t. t.

MATLAB Code

```
%.....Program to decompose a given audio signal using Discrete Wavelet Transform....%
N=3;  %Three level decomposition
x=wavread('song.wav');
plot(x);
[C,L]=wavedec(x,'wavelet',N); % 'wavelet' should be replaced with 'haar','db2', etc.
% C contains the approximation and detail coefficients.
% L contains the length of the respective approx. and detailed coefficients

%....Program to reconstruct the audio signal using Discrete Wavelet Transform....%
z=waverec(C, L,'wavelet'); % 'wavelet' should be replaced with 'haar','db2', etc.
plot(z);
```

2.5 Solved Problems

Problem How many levels are required for decomposing the signal shown in Fig. 2.13 using Haar wavelet? Find the approximation and detailed coefficients, c_k and $d_{j,k}$, in different levels.

Solution Using Eq. 2.21,

$$c_k = [1, 0, 2, 3, -1]$$

Using Eq. 2.22,

$$d_{0,k} = \delta[n-5]$$

Further,

$$d_{1,k} = d_{2,k} = d_{3,k} = d_{4,k}, \ldots = 0$$

Problem In Fig. 2.6, what is the best suited window width for different durations?

Solution The optimum window width will be that which gives the best time-frequency resolution. The best time-frequency resolution will be obtained from 0 to 200 units of time by using a window of duration 100 units. The optimum window width for the duration from 200 to 400 units of time will be 20 units, and the optimum window width for the duration from 400 to 600 units of time will be 200 units.

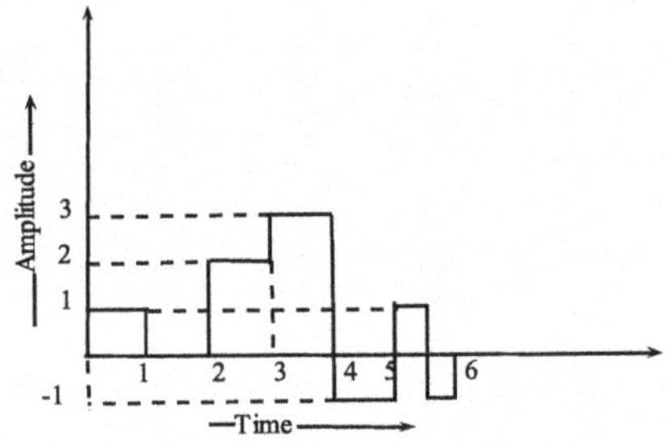

Fig. 2.13 A sample signal

Chapter 3
Multiresolution Analysis

Depending on whether the signal $f(t)$ is represented using $\psi(t)$, $\psi(2t)$, $\psi(4t)$, so on, and their shifted versions, the time resolution of the representation increases, while the frequency resolution decreases. Thus we get multiresolution by changing the scale and shifts of the wavelet function, $\psi(t)$.

3.1 Heisenberg's Uncertainity Principle

Heisenberg's uncertainity principle in wavelet theory states that if the standard deviation of a wavelet is σ_t and the standard deviation of its frequency domain wave, represented by $\Phi(\omega)$, is σ_ω, then

$$\sigma_t \times \sigma_\omega \leq \frac{1}{2} \tag{3.1}$$

Figure 3.1 shows the variation in time and frequency resolution for different wavelets having different σ_t and σ_ω. The time and frequency resolution for wavelet transform is as shown in Fig. 3.2a. As the scaling factor increases, the frequency resolution increases, while the time resolution decreases. If we consider two tiles in Fig. 3.2a, it is obvious that the time resolution is higher for tile *a* compared to tile *b*, while the frequency resolution is lower for tile *a* compared to tile *b*.

Figure 3.2b represents the time-frequency tiling of sampled signal, and Fig. 3.2c represents the time-frequency tiling of STFT. From Fig. 3.2c, it is clear that the time-frequency resolution is fixed for STFT for a particular window size. If the window size is decreased, the time resolution increases and frequency resolution decreases.

M. S. Sinith et al., *Wavelets and Multi-Resolution Analysis*,
https://doi.org/10.1007/978-3-031-92750-8_3

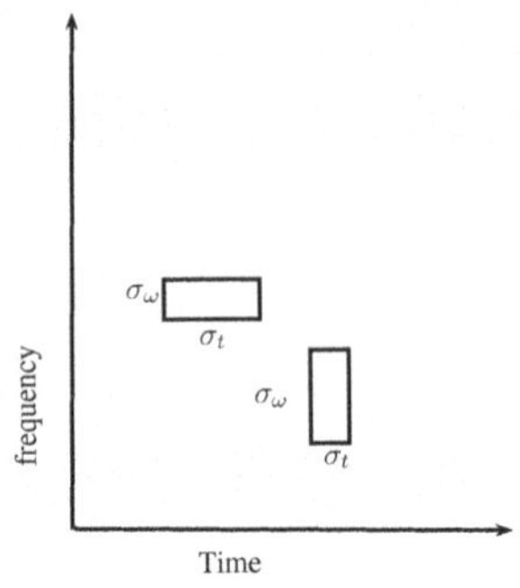

Fig. 3.1 Time-frequency resolution

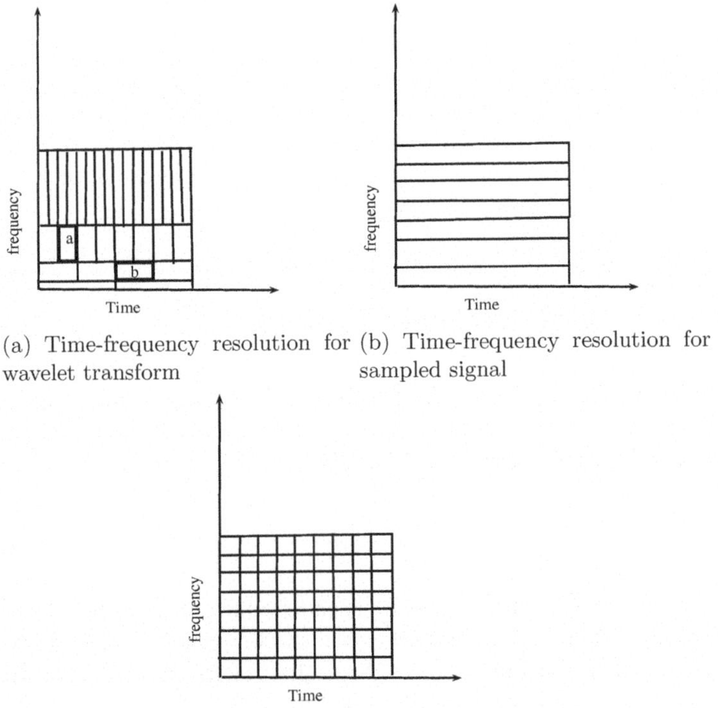

(a) Time-frequency resolution for wavelet transform

(b) Time-frequency resolution for sampled signal

(c) Time-frequency resolution for STFT

Fig. 3.2 Time-frequency resolution

3.2 Fundamentals of Multiresolution Analysis (MRA)

Multiresolution analysis (MRA) is a mathematical framework used to systematically construct and analyze wavelets and their associated scaling functions. It enables the decomposition of a signal into a hierarchy of approximations and details at different scales or resolutions, allowing for the analysis of signal features at various levels of detail.

The key components of MRA include:

- **Nested subspaces**: Represent different levels of approximation.
- **Scaling functions**: Define the coarse approximation of a signal.
- **Wavelet functions**: Capture the detailed information lost in approximations.

MRA forms the theoretical foundation for constructing **orthogonal** and **biorthogonal wavelet bases**, which are widely applied in signal processing, such as image compression, and numerical analysis.

3.3 Properties of MRA

Nesting Property

Let V_j be a sequence of subspaces of functions in $L^2(\mathbb{R})$, such that:

$$V_j \subset L^2(\mathbb{R}), \quad \forall j \in \mathbb{Z} \tag{3.2}$$

These subspaces satisfy the **nesting condition**, forming a hierarchical structure:

$$\cdots \subset V_{j-1} \subset V_j \subset V_{j+1} \subset \cdots \tag{3.3}$$

This ensures that each subspace V_j is contained in the next higher-resolution space V_{j+1}, meaning:

$$V_j \subset V_{j+1}, \quad \forall j \in \mathbb{Z} \tag{3.4}$$

As a result, functions in V_j can be approximated with greater precision in V_{j+1}, allowing for a progressively finer representation of signals in $L^2(\mathbb{R})$.

Closure Property

The closure property of an MRA ensures that the union of all subspaces spans the entire space $L^2(\mathbb{R})$. This means any function in $L^2(\mathbb{R})$ can be approximated arbitrarily well using the functions in these subspaces. Mathematically, this is expressed as

$$\overline{\bigcup_{j \in \mathbb{Z}} V_j} = L^2(\mathbb{R}) \tag{3.5}$$

This guarantees that for any function $f(t) \in L^2(\mathbb{R})$, there exists a sequence of approximations in V_j that converges to $f(t)$. As the resolution level j increases (i.e., as $j \to \infty$), the space V_j becomes finer and more detailed, capturing more variations of the function. In the limit, the union of all these spaces spans the entire function space $L^2(\mathbb{R})$, ensuring that no part of the signal is lost in the decomposition.

Shrinking Property

As the resolution index j decreases, the space V_j approximates functions with coarser resolution. This means that lower-scale subspaces V_j capture fewer details of the signal.

$$\bigcap_{j \in \mathbb{Z}} V_j = \{0\} \tag{3.6}$$

This implies that the only function contained in all subspaces is the trivial function $f(t) = 0$, meaning the representation loses all details at extremely coarse resolutions.

Scaling Property

The **scaling property** states that if a function $f(t)$ belongs to the space V_j, then its scaled version should belong to the base space V_0. That is, for any function:

$$f(t) \in V_j \quad \Rightarrow \quad f(2^{-j}t) \in V_0 \tag{3.7}$$

This ensures that the subspaces V_j can be generated from a single space V_0 through a scaling operation.

The scaling function $\phi(t)$, which generates these subspaces, satisfies the relation:

$$\phi(t) \in V_0 \quad \Rightarrow \quad \phi(2^{-j}t) \in V_j \tag{3.8}$$

Thus, scaling plays a fundamental role in the hierarchical construction of an MRA.

Shift Property

The shift property states that the subspaces V_j are invariant under integer translations. This means that if a function $f(t)$ belongs to V_j, then its shifted version should also belong to V_j:

$$f(t) \in V_j \quad \Rightarrow \quad f(t-k) \in V_j, \quad \forall k \in \mathbb{Z} \tag{3.9}$$

In particular, the **scaling function** $\phi(t)$ satisfies

$$\phi(t) \in V_0 \quad \Rightarrow \quad \phi(t-k) \in V_0, \quad \forall k \in \mathbb{Z} \tag{3.10}$$

Orthonormality Condition

The orthonormality condition ensures that the translated basis functions form an orthonormal set. Specifically, the functions $\phi(t-k)$ should satisfy:

$$\langle \phi(t-k), \phi(t-m) \rangle = \delta_{k,m}, \quad \forall k, m \in \mathbb{Z} \tag{3.11}$$

where $\delta_{k,m}$ is the Kronecker delta, defined as

$$\delta_{k,m} = \begin{cases} 1, & k = m, \\ 0, & k \neq m \end{cases} \tag{3.12}$$

This condition ensures that the set of translated scaling functions forms an orthonormal basis for V_0, which extends to other subspaces V_j through scaling.

3.4 Approximation and Detail Coefficients

Let us represent the space spanned by $\phi(t)$ and its shifted versions as V_0 (Fig. 3.3). Any arbitrary signal, $f(t)$, can be approximated in the space V_0 as

$$f_0(t) = \sum_{k=0}^{\infty} c_k \phi(t-k) \tag{3.13}$$

Let $\psi(t)$ and its shifted versions span the space represented by W_0 and let $\psi(2t)$ and its shifted versions span the space represented by W_1. A better approximation of the signal $f(t)$ can be obtained by adding the next level details (details or components of the signal $f(t)$ in W_0). Thus,

$$f_1(t) = \sum_{k=0}^{\infty} c_k \phi(t-k) + \sum_{k=0}^{\infty} 2^{\frac{1}{2}} d_{1,k} \psi(2t-k) \tag{3.14}$$

where $f_1(t)$ is the approximation of the signal in the approximation space V_1. Thus using Eqs. 3.13 and 3.14,

$$V_1 = V_0 \oplus W_0 \tag{3.15}$$

Since all the signals in V_0 and W_0 are orthogonal to each other,

$$V_0 \perp W_0 \tag{3.16}$$

It is also clear that

$$V_0 \subset V_1 \tag{3.17}$$

$$W_0 \subset V_1 \tag{3.18}$$

Similar to Eq. 3.15,

$$V_2 = V_1 \oplus W_1 \tag{3.19}$$

Substituting Eq. 3.15 in Eq. 3.19,

$$V_2 = W_1 \oplus W_0 \oplus V_0 \tag{3.20}$$

Thus the approximation of the signal $f(t)$ at V_2 will be

$$f_2(t) = \sum_{k=0}^{\infty} c_k \phi(t-k) + \sum_{k=0}^{\infty} \sqrt{2} d_{1,k} \psi(2t-k) + \sum_{k=0}^{\infty} 2 d_{2,k} \psi(4t-k) \tag{3.21}$$

Since $V_0 \perp W_0 \perp W_1$, graphical representation of Eq. 3.20 is as shown in Fig. 3.3. Generalizing Eq. 3.21, the approximation of the signal $f(t)$ at the approximation level P can be represented as

$$f_P(t) = \sum_{k=0}^{\infty} c_k \phi(t-k) + \sum_{j=0}^{P-1} \sum_{k=0}^{\infty} 2^{\frac{j}{2}} d_{j,k} \psi(2^j t - k) \tag{3.22}$$

Let $f(t)$ be a signal as shown in Fig. 3.4a which is a sinusoidal wave with an impulse added at $t = 200\ units$. The approximations of this signal at different levels, using Haar wavelet, are shown in Figs. 3.4b–e. It can be easily understood that higher frequency signals are excluded in lower levels of approximation. For example in Fig. 3.4b, the impulse is not present.

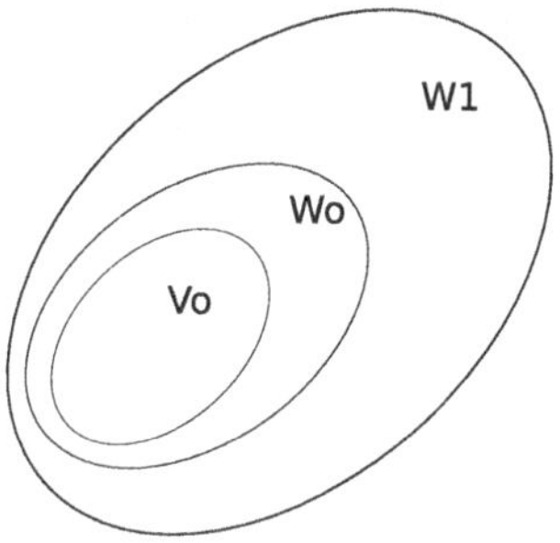

Fig. 3.3 Venn diagram representation

Thus, as P approaches ∞,

$$f(t) = f_{\infty}(t) = \sum_{k=0}^{\infty} c_k \phi(t-k) + \sum_{j=0}^{\infty} \sum_{k=0}^{\infty} 2^{\frac{j}{2}} d_{j,k} \psi(2^j t - k) \qquad (3.23)$$

Hence,

$$V_{-\infty} \subset \ldots \subset V_0 \subset V_1 \subset V_2 \subset V_3 \subset \ldots \subset V_{\infty} \qquad (3.24)$$

where approximation of the signal $f(t)$ in the space $V_{-\infty}$ denotes the DC value of the signal.

3.5 Subband Coding and Multiresolution Analysis

Subband coding is a fundamental technique in signal processing that decomposes a signal into different frequency bands for efficient representation and analysis. This technique is closely linked with multiresolution analysis (MRA), which systematically analyzes signals at multiple scales. The discrete wavelet transform (DWT) is a key tool in MRA, enabling hierarchical signal decomposition into approximation and detail components.

When a signal is passed through a half-band low-pass filter, it retains only the lower frequency components while discarding the higher frequencies. According to Nyquist's rule, half of the samples can be eliminated without losing any information, since the highest frequency in the signal is reduced from π radians to $\pi/2$ radians. This process, known as downsampling, reduces the number of samples by half, effectively doubling the scale of the signal. Filtering and downsampling affect two key properties of the signal:

- Scale: The low-pass filtering step does not alter the scale, but downsampling changes it by reducing the number of samples.

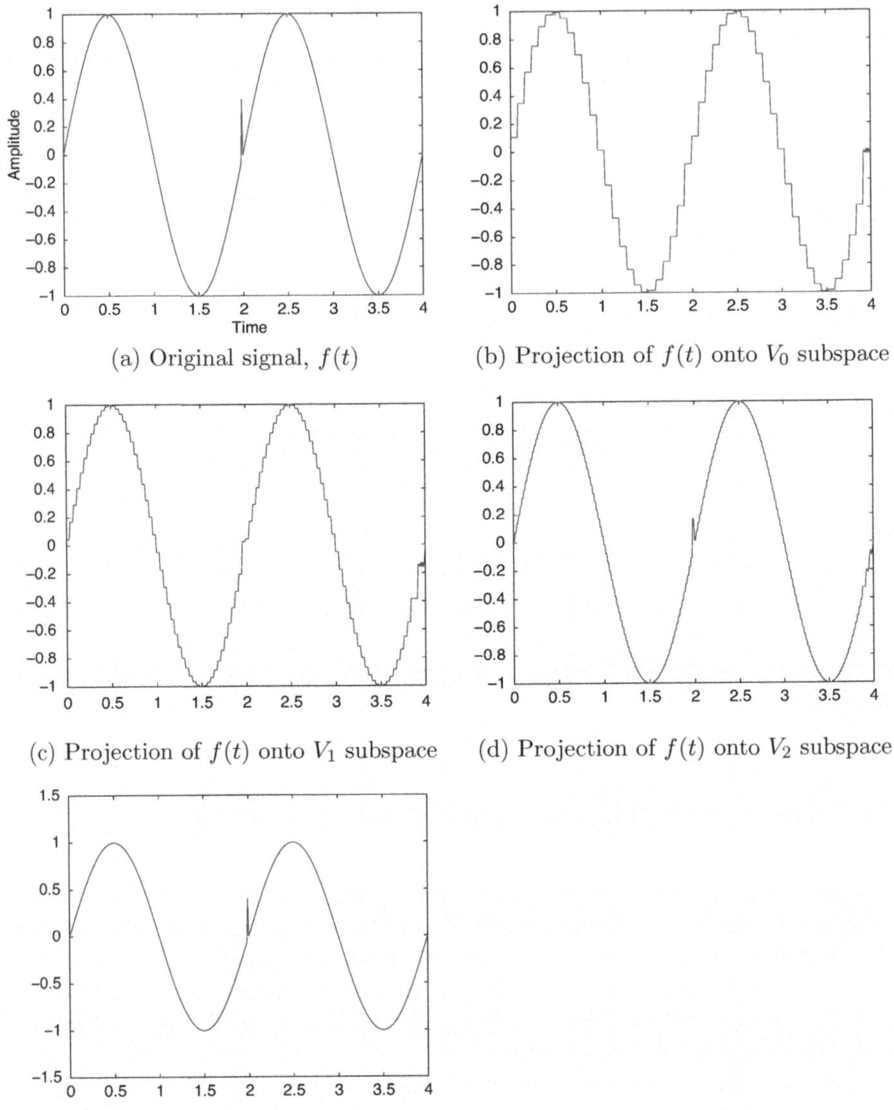

(a) Original signal, $f(t)$

(b) Projection of $f(t)$ onto V_0 subspace

(c) Projection of $f(t)$ onto V_1 subspace

(d) Projection of $f(t)$ onto V_2 subspace

(e) Projection of $f(t)$ onto V_3 subspace

Fig. 3.4 Projection of $f(t)$ onto different subspaces

- Resolution: The resolution is determined by the amount of information in the signal. Since low-pass filtering removes half of the spectral components, the resolution is effectively halved. However, subsampling does not impact the resolution, as it only removes redundant samples.

The signal $x[n]$ is first filtered using a low-pass filter $h[n]$ and a high-pass filter $g[n]$:

$$y_{\text{low}}[n] = \sum_k h[k] \cdot x[n-k] \tag{3.25}$$

$$y_{\text{high}}[n] = \sum_k g[k] \cdot x[n-k] \tag{3.26}$$

The filtered signals are then downsampled by 2:

$$x_{\text{low}}[n] = y_{\text{low}}[2n], \quad x_{\text{high}}[n] = y_{\text{high}}[2n] \tag{3.27}$$

This operation removes redundant samples while preserving essential frequency information. The DWT decomposes a signal into different frequency bands using successive high-pass and low-pass filtering. In the first level of decomposition, the original signal is filtered by both the high-pass and low-pass filters and then downsampled by 2. This process results in two outputs:

- The low-pass filter output retains the lower half of the frequency spectrum and has half the number of samples.
- The high-pass filter output captures the higher frequency details, also with half the number of samples.

Each subsequent decomposition level applies the same filtering and subsampling process to the low-pass-filtered signal, further breaking it down into lower and lower frequency components. As a result, at each level:

- Time resolution is halved, as the number of samples decreases.
- Frequency resolution is doubled, as the bandwidth is reduced.

The subsequent decomposition process is illustrated in Fig. 3.5.

Example: Audio Signal with Sampling Frequency f_s and 256 Sample Points
In digital signal processing, the highest frequency that can be represented in a sampled signal is the Nyquist frequency, which is half the sampling frequency f_s. This frequency corresponds to π radians per sample in the normalized angular frequency domain. Therefore, the frequency range of $[0, \pi]$ radians/sample can be equivalently expressed in terms of physical frequency as $[0, f_s/2]$ Hz. The frequency bands at each decomposition level of the discrete wavelet transform (DWT) can thus be represented in terms of the sampling frequency instead of angular frequency.

First Level Decomposition At the first level of decomposition, the signal is passed through a high-pass and a low-pass filter, followed by subsampling:

- The high-pass filter output contains 128 samples, spanning the frequency range $[f_s/4, f_s/2]$ Hz. Since half of the frequency band is removed, the frequency resolution is doubled.
- The low-pass filter output contains 128 samples, spanning the range $[0, f_s/4]$ Hz.

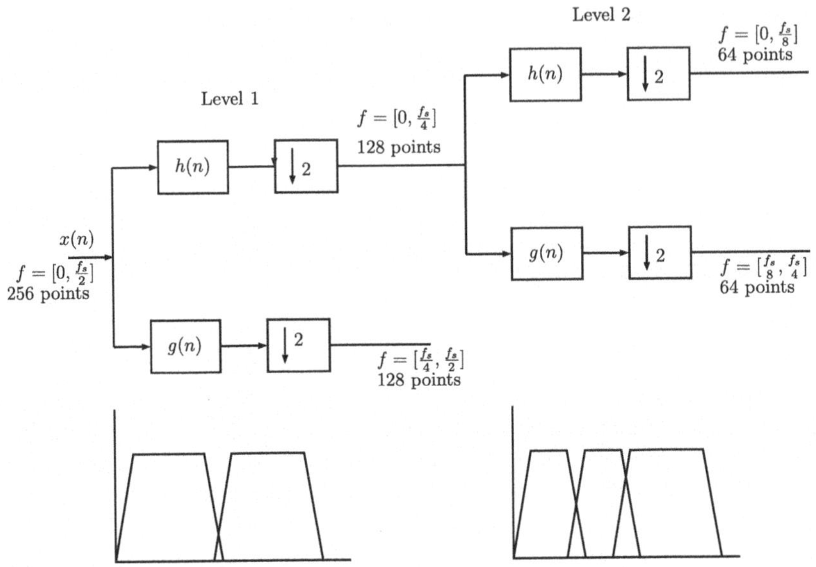

Fig. 3.5 Figure showing subsequent decomposition

Second-Level Decomposition The low-pass-filtered signal undergoes further decomposition:

- The second high-pass filter output contains 64 samples, covering the frequency range $[f_s/8, f_s/4]$ Hz.
- The second low-pass filter output contains 64 samples, spanning $[0, f_s/8]$ Hz.

At this stage:

- Time resolution has decreased by a factor of 4 (from 256 to 64 samples).
- Frequency resolution has increased by a factor of 4, since each band now occupies only one-fourth of the original frequency range.

Subsequent Decomposition Levels This process continues, where each low-pass-filtered output undergoes further decomposition:

- Third level: 32 samples, spanning $[0, f_s/16]$ Hz and $[f_s/16, f_s/8]$ Hz.
- Fourth level: 16 samples, spanning $[0, f_s/32]$ Hz and $[f_s/32, f_s/16]$ Hz.
- Fifth level: 8 samples, spanning $[0, f_s/64]$ Hz and $[f_s/64, f_s/32]$ Hz.
- Sixth level: 4 samples, spanning $[0, f_s/128]$ Hz and $[f_s/128, f_s/64]$ Hz.
- Seventh level: 2 samples, spanning $[0, f_s/256]$ Hz and $[f_s/256, f_s/128]$ Hz.

For this example, there are seven levels of decomposition, each reducing the number of samples by half compared to the previous level. The final DWT representation is obtained by concatenating all detail and approximation coefficients from the last level to the first level. Since no information is lost in the filtering and subsampling process, the total number of DWT coefficients remains 256, the same as the original signal.

3.6 Fast Wavelet Transform

For Haar wavelet, we have already seen that

$$\phi(t) = \phi(2t) + \phi(2t-1)$$
$$\psi(t) = \phi(2t) - \phi(2t-1)$$

which can be rewritten as

$$\phi(t) = h(0)\phi(2t) + h(1)\phi(2t-1) \tag{3.28}$$
$$\psi(t) = g(0)\phi(2t) + g(1)\phi(2t-1) \tag{3.29}$$

where $h(n) = [1, 1]$ and $g(n) = [1, -1]$. For any wavelet, Eqs. 3.28 and 3.29 can be represented more generally as

$$\phi(t) = \sum_{k=0}^{N} h(k)\phi(2t-k) \tag{3.30}$$
$$\psi(t) = \sum_{k=0}^{N} g(k)\phi(2t-k) \tag{3.31}$$

substituting $t = (2^j t - k)$ in Eqs. 3.30 and 3.31

$$\phi(2^j t - k) = \sum_{k=0}^{N} h(k)\phi(2(2^j t - k) - k) \tag{3.32}$$
$$\psi(2^j t - k) = \sum_{k=0}^{N} g(k)\phi(2(2^j t - k) - k) \tag{3.33}$$

Multiplying these equations with the signal to be analyzed, $f(t)$, and integrating:

$$\int f(t)\phi(2^j t - k)dt = \int f(t)\sum_{k=0}^{N} h(k)\phi(2^{j+1}t - 3k)dt \tag{3.34}$$

$$\int f(t)\psi(2^j t - k)dt = \int f(t)\sum_{k=0}^{N} g(k)\psi(2^{j+1}t - 3k)dt \tag{3.35}$$

i.e.,

$$\int f(t)\phi(2^j t - k)dt = \int f(t)\sum_{k=0}^{N} h(k)\phi((2^{j+1}t - k) - 2k)dt \tag{3.36}$$

$$\int f(t)\psi(2^j t - k)dt = \int f(t)\sum_{k=0}^{N} g(k)\psi((2^{j+1}t - k) - 2k)dt \tag{3.37}$$

Using Eqs. 2.21 and 2.22 in Eqs. 3.36 and 3.37 yields

$$c_{j,k} = \sum_{k=0}^{N} h(k)c_{j+1,2k} \tag{3.38}$$

$$d_{j,k} = \sum_{k=0}^{N} g(k)c_{j+1,2k} \tag{3.39}$$

Hence, $c_{j,k}$ can be obtained by passing it through a filter having impulse response $h(-n)$ followed by downsampling. Similarly, $d_{j,k}$ can be obtained by passing it through a filter having impulse response $g(-n)$ followed by downsampling. This algorithm using which the approximation and detailed coefficients at a lower level can be found using approximation level at a higher level is called fast wavelet transform.

3.7 Solved Problems

Problem Given two properties of classical wavelet,

$$\sum h^2(n) = 2 \tag{3.40}$$

$$\sum h(n) = 2 \tag{3.41}$$

1. Prove that Eq. 3.44 implies that the shifted versions of the scaling function is orthogonal to each other.
2. Prove that Eq. 3.45 implies that the area under the curve $\phi(t)$ is one.

Solution Squaring and integrating Eq. 3.30 w.r.t. the variable t,

$$\int_0^\infty \phi^2(t)\,dt = \int_0^\infty [\sum_{k=0}^{N} h(k)\phi(2t-k)]^2\,dt$$

i.e.,

$$\int_0^\infty \phi^2(t)\,dt = \int_0^\infty (h(0)\phi(2t) + h(1)\phi(2t-1) + ... + h(N)\phi(2t-N))^2. \tag{3.42}$$

Equation 2.16 can also be written as

$$\int_0^\infty \phi(2t-k)\phi(2t-l)dt = \delta(k-l)\ \forall\, k, l \in Z \tag{3.43}$$

Using Eqs. 3.42, 3.43 and the fact that $\int \phi^2(2t-k)\,dt = 0$,

$$\sum h^2(n) = 2 \tag{3.44}$$

Proceeding in the same manner, it can be proved that

$$\sum h(n) = 2 \tag{3.45}$$

Problem If V_j and W_j represent the approximation and detail subspace at level j, state and justify the following statements:

1. $V_0 \perp V_1 \perp V_2$
2. $V_0 \perp W_0 \perp W_1$
3. $V_3 = W_2 \oplus W_1 \oplus W_0 \oplus V_0$

Solution

1. From Eq. 3.24, it is clear that $V_0 \subset V_1 \subset V_2$. Hence, the statement $V_0 \perp V_1 \perp V_2$ is false.
2. From Fig. 3.3, it is clear that $V_0 \perp W_0 \perp W_1$.
3. From Eq. 3.20, $V_2 = W_1 \oplus W_0 \oplus V_0$. Extending this to one more level, $V_3 = W_2 \oplus W_1 \oplus W_0 \oplus V_0$.

Chapter 4
Discrete Wavelet Transform and Filter Bank Theory

We know a wavelet can be completely characterized by its scaling and wavelet function. But in many applications, the approximation and detail coefficients, i.e., $h[n]$ and $g[n]$, are easy to work with. We will now see that the refinement relation involves these coefficients only. This chapter relates the wavelet transform with filter banks. Filter banks constitute an important part of multirate signal processing. The implementation of wavelet transform with the help of filter banks made the design and analysis of wavelets easier. So it is necessary to understand how the wavelet transform is related with the filter bank structure.

4.1 QMF Filter Bank

A QMF filter bank is a two channel filter bank consisting of filters, decimators, and interpolators. We will discuss each block in detail.

Filter Bank

Filter bank, as its word meaning indicates, is a collection of filters. But the speciality is that the filters either have a common input or a common output. Filter banks are classified into two, analysis filter bank and synthesis filter bank.

An analysis filter bank is shown in Fig. 4.1a. It has a set of filters arranged in parallel with a common input signal. The input signal $x[n]$ is divided into M different signals, known as subband signals depending on the frequency response of the filters, i.e., the output of each branch is different depending upon the response of the filter in the corresponding branch. A synthesis filter bank is shown in Fig. 4.1b. The input signal to each branch is different, and the output is the sum of the outputs

M. S. Sinith et al., *Wavelets and Multi-Resolution Analysis*,
https://doi.org/10.1007/978-3-031-92750-8_4

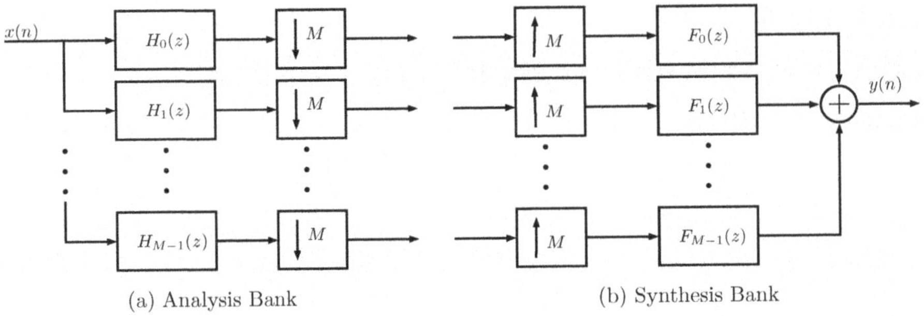

Fig. 4.1 Types of filter banks

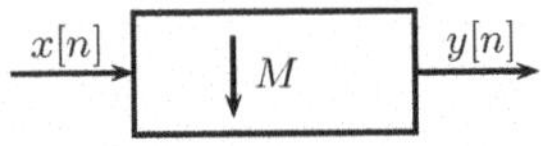

Fig. 4.2 Decimator

of different filters, i.e., the filter bank combines the M subband signals into a single signal $y[n]$.

The bandwidth of each subband signal from each individual filter is different from the bandwidth of the original signal. For example, if $H_0(z)$ is a low-pass filter with cutoff frequency $\frac{\pi}{2}$, then the output of this filter will consist of frequencies from 0 to $\frac{\pi}{2}$ only. A low-frequency signal has slow variations in the time domain. Hence, the output signal from $H_0(z)$ need not be sampled at the sampling rate of input signal $x(n)$. Hence, while coming into practical applications, we can reduce the sampling frequency. The process of changing sampling rates is discussed in the next section.

Decimation

The process of reducing the sampling rate is called decimation. The symbol of an M-fold decimator is shown in Fig. 4.2. Here M indicates the factor by which the signal is downsampled. Note that M is always a positive integer. It reduces the sampling rate by choosing every Mth sample in the original signal. In other words, a decimator chooses only those samples of $x(n)$ which occur at multiples of M and produces the output $y(n)$. A decimator is also called a downsampler or a subsampler or a compressor.

We can understand the working of a decimator using an example. The original signal $x[n]$ is shown in Fig. 4.3a. Assume that the decimation factor M is 2. So the decimator retains every second sample and all other samples are discarded. The output is shown in Fig. 4.3b.

From the example, we can say that a decimation process is composed of two steps. First, it removes the samples in some locations and then compresses the

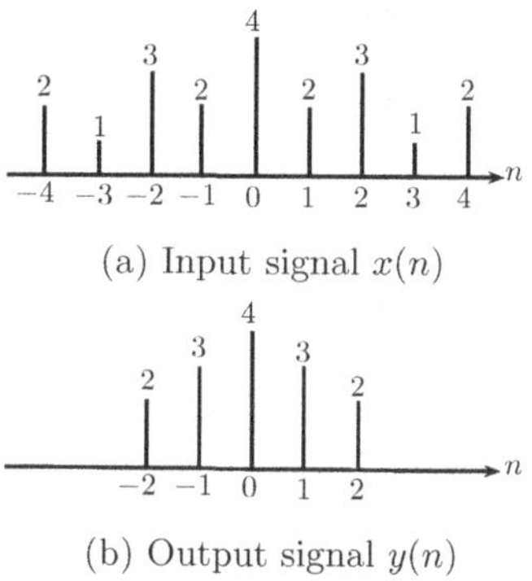

Fig. 4.3 Example showing decimation process

signal. More clearly, in the above case, the decimator first replaces the values at odd locations with zeros. Then it compresses this signal by a factor of 2 to get $y[n]$. Also note that the removal of some samples from the input makes the process of decimation irreversible.

Time Domain Representation

It is mentioned that an M-fold decimator retains every Mth sample in the input signal $x[n]$. Hence, an M-fold decimator with an input signal $x[n]$ and output signal $y[n]$ can be represented in the time domain as

$$y[n] = x[Mn] \tag{4.1}$$

If the time period of the input signal is T, then the input sampling frequency F is given by $F = \frac{1}{T}$. For the output signal, the sampling frequency F' is given by $F' = \frac{1}{MT} = \frac{F}{M}$. Hence, the sampling rate of the output signal is $\left(\frac{1}{M}\right)$th of the input signal sampling rate.

Frequency Domain Representation

We cannot directly represent the Z transform of the decimated signal in terms of the input signal, because while doing decimation, we discarded some samples from the input. So for the purpose, consider an intermediate sequence as given below.

$$x_1[n] = \begin{cases} x[n], & \text{n} = 0, \pm M, \pm 2M.......... \\ 0, & \text{otherwise} \end{cases} \tag{4.2}$$

The signal $x_1[n]$ can be written in terms of $x[n]$ as $x_1[n] = c[n].x[n]$ where $c[n]$ is known as the comb sequence and is given by

$$c[n] = \begin{cases} 1, & \text{n} = 0, \pm M, \pm 2M.......... \\ 0, & \text{otherwise} \end{cases} \tag{4.3}$$

$c[n]$ can also be expressed as $\frac{1}{M}\sum_{k=0}^{M-1} W_M^{kn}$ where $W_M = e^{\frac{-j2\Pi}{M}}$.
So the z transform of output signal can be written as

$$\begin{aligned} Y[z] &= \sum_{n=-\infty}^{\infty} y[n]z^{-n} \\ &= \sum_{n=-\infty}^{\infty} x[Mn]z^{-n} \\ &= \sum_{n=-\infty}^{\infty} x_1[Mn]z^{-n} \\ &= \sum_{k=-\infty}^{\infty} x_1[k]z^{-k/M} \quad \text{putting k=Mn, so that n=}\frac{k}{M} \\ &= X_1\left[z^{\frac{1}{M}}\right] \end{aligned} \tag{4.4}$$

Therefore, we can write

$$\begin{aligned} Y[z] &= X_1(z^{\frac{1}{M}}) \\ &= \sum_{n=-\infty}^{\infty} c[n]x[n]z^{-\frac{n}{M}} \\ &= \frac{1}{M}\sum_{n=-\infty}^{\infty}\left(\sum_{k=0}^{M-1} W_M^{kn}\right)x[n]z^{-\frac{n}{M}} \\ &= \frac{1}{M}\sum_{k=0}^{M-1}\left(\sum_{n=-\infty}^{\infty} x[n]W_M^{kn}\right)z^{-\frac{n}{M}} \\ &= \frac{1}{M}\sum_{k=0}^{M-1} X\left[z^{\frac{1}{M}}W_M^{-k}\right] \end{aligned} \tag{4.5}$$

From Eq. 4.5, we can easily write the expression for $Y(e^{j\omega})$.

$$Y(e^{j\omega}) = \frac{1}{M}\sum_{k=0}^{M-1} X\left(e^{\frac{j(\omega+2\Pi k)}{M}}\right) \tag{4.6}$$

Let us have a graphical representation of the decimation process. Consider a signal $x[n]$ whose Fourier transform is given in Fig. 4.4a.

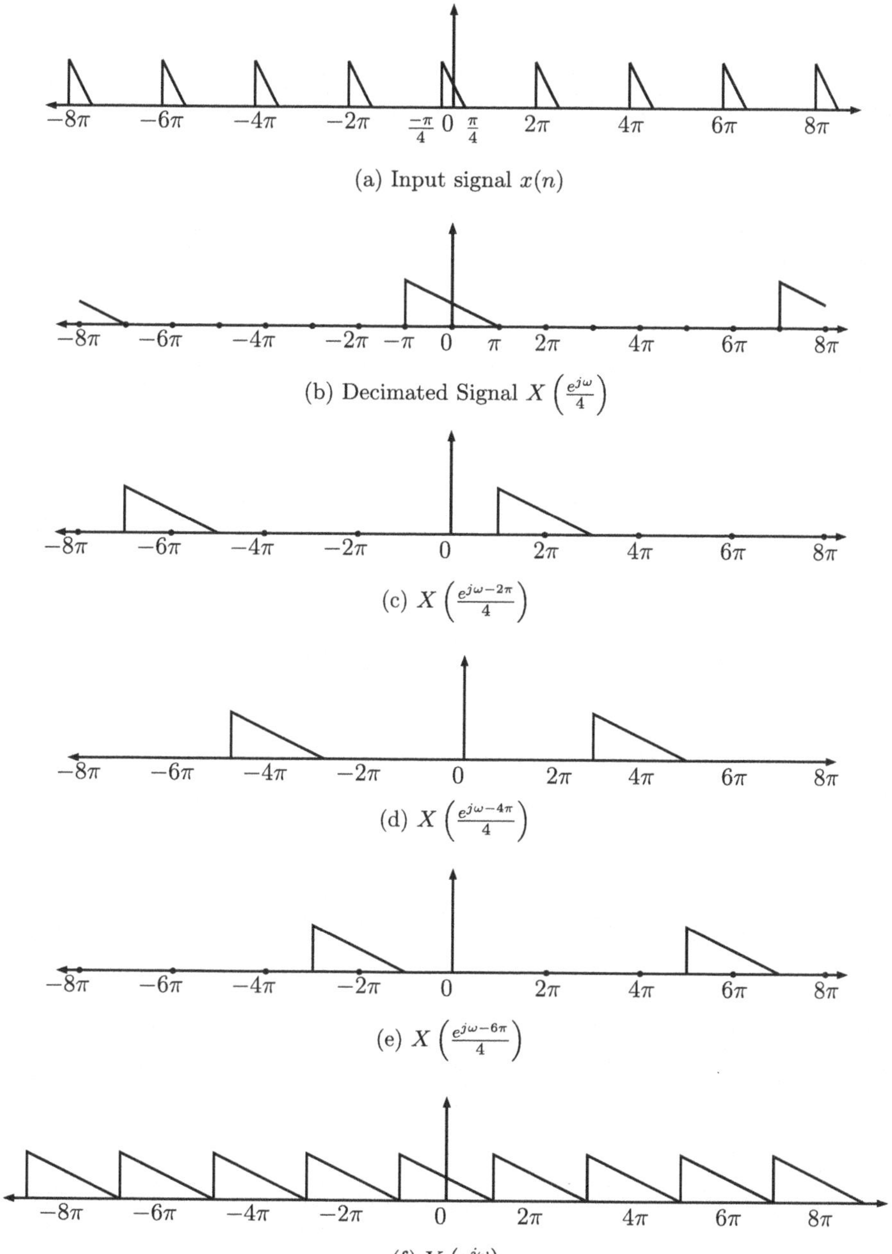

(a) Input signal $x(n)$

(b) Decimated Signal $X\left(\frac{e^{j\omega}}{4}\right)$

(c) $X\left(\frac{e^{j\omega-2\pi}}{4}\right)$

(d) $X\left(\frac{e^{j\omega-4\pi}}{4}\right)$

(e) $X\left(\frac{e^{j\omega-6\pi}}{4}\right)$

(f) $Y\left(e^{j\omega}\right)$

Fig. 4.4 Decimation process in frequency domain

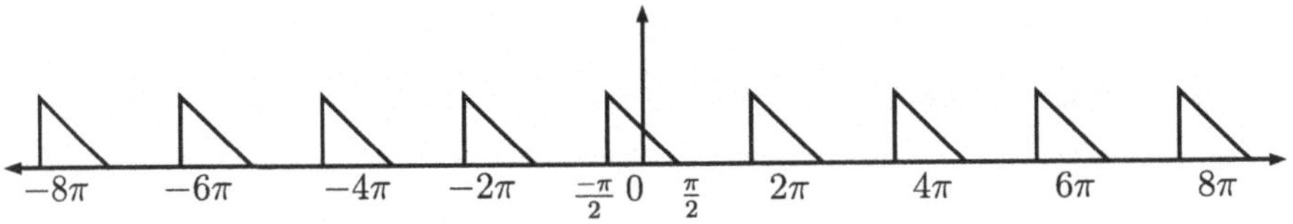

(a) Input signal $x(n)$

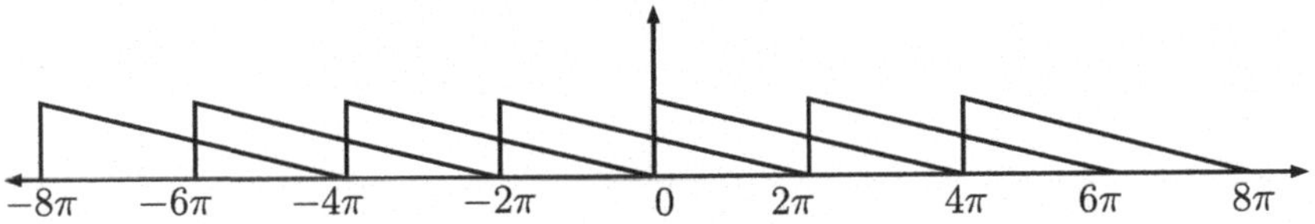

(b) $Y(e^{j\omega})$

Fig. 4.5 Decimation process in frequency domain with aliasing

The Fourier transform of the output, i.e., $Y(e^{j\omega})$, can be derived from Eq. 4.6. As per the equation, $Y(e^{j\omega})$ can be obtained by doing the following steps:

1. Expand $X(e^{j\omega})$ by the decimation factor, M.
2. Create a copy of this expanded signal by shifting it by 2π.
3. Create $M-1$ copies of such expansions by shifting the signal in multiples of 2π.
4. Add all outputs of steps 2 and 3. Then divide the result by M.

For demonstration purpose, take $M = 4$. The output $Y(e^{j\omega})$ can be drawn by successively following the above steps:

1. The input signal $X(e^{j\omega})$ is expanded by the decimation factor 4 and is shown in Fig. 4.4b.
2. This expanded signal is shifted by 2π and is shown in Fig. 4.4c.
3. The signal is shifted in 4π and 6π to create $M-1=3$ copies of the input signal which is shown in Figs. 4.4d and e.
4. The signals from steps 2 and 3 are added and divided by M to obtain the final output as in Fig. 4.4f.

We have to derive some interesting inferences from the above figures. On careful examination, we can see that our input signal has a bandwidth of $\pi/2$ (it exists from $(-\pi/4)$ to $(+\pi/4)$). The output signal shown in Fig. 4.4f is a nonoverlapped one. Now assume the input signal extends from $(-\pi/2)$ to $(+\pi/2)$ as shown in Fig. 4.5a. Upon downsampling this signal by a factor of 4, the output signal will be as shown in Fig. 4.5b (which can be easily derived by following the above steps). Here as you can see, the signal is overlapping, and this process is known as *aliasing*. In general, to avoid aliasing, the input signal should be bandlimited to the region $|\omega| < \pi/M$.

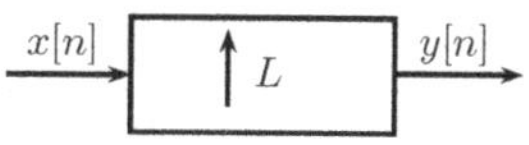

Fig. 4.6 Interpolator

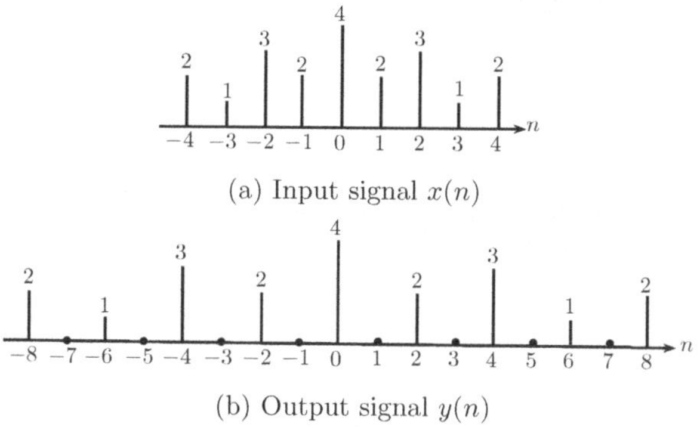

Fig. 4.7 Example showing interpolation process

Interpolation

The process of increasing the sampling rate is called interpolation. An L-fold interpolator is shown in Fig. 4.6. It increases the sampling rate by inserting $(L - 1)$ zeros between two samples of the input signal. An interpolator is also called an upsampler or an expander.

To understand the operation, let us consider an example. The original signal $x[n]$ is shown in Fig. 4.7a, which is same as that we took in decimator example. Assume that the interpolation factor L is 2. So the interpolator inserts $(L - 1) = 1$, zero sample between two samples. The output is shown in Fig. 4.7b. Actually, this process is called upsampling. The interpolator performs the process of upsampling, and additionally, it replaces zero values by nonzero values, which is done with the help of a filter. But usually, the terms upsampling and interpolation are used interchangebly.

Time Domain Representation

An upsampler increases the sampling rate of the incoming signal. If the input signal is $x[n]$ and the output signal is $y[n]$, the expression for the output signal is given by

$$y[n] = x\left[\frac{n}{L}\right] \tag{4.7}$$

To understand the concept that an upsampler increases the sampling frequency, assume that the sampling period of the input signal is T. Then the sampling period of the output signal can be written as $\frac{T}{L}$. The sampling frequency of input signal $F = \frac{1}{T}$ and the sampling frequency of output, $F' = \frac{L}{T} = LF$, which is L times more than the sampling frequency of input. Thus our requirement to increase the sampling rate is achieved.

Frequency Domain Representation
The process of finding the Z transform of an upsampler is simpler than that of a downsampler. Let the input signal be $x[n]$, the output signal be $y[n]$, and L be the upsampling factor. Then the Z transform of output is given by

$$\begin{aligned} Y[z] &= \sum_{n=-\infty}^{\infty} y[n]z^{-n} \\ &= \sum_{n=-\infty}^{\infty} x\left[\frac{n}{L}\right]z^{-n} \\ &= \sum_{m=-\infty}^{\infty} x[m]z^{-LM} \qquad \text{by putting} \quad m = \frac{n}{L} \\ &= X[z^L] \end{aligned} \tag{4.8}$$

Putting $z = e^{j\omega}$, the above equation can be written as

$$Y(e^{j\omega}) = X(e^{j\omega L}) \tag{4.9}$$

As in the case of decimator, we now go for the graphical representation of an interpolation process. Take the same signal $X(e^{j\omega})$ as we had seen in decimation as shown in Fig. 4.8a. (It is the same signal. Only the scale of drawing is changed!) Assume the interpolation factor L to be 4. From Eq. 4.9, we know the output $Y(e^{j\omega})$

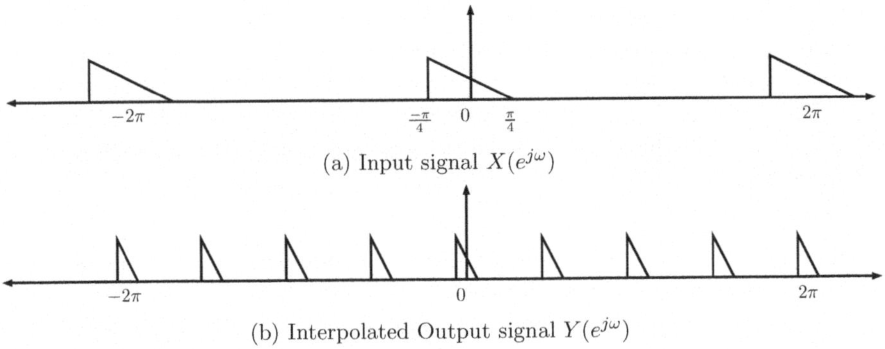

Fig. 4.8 Interpolation process in frequency domain

is a compressed version of $X(e^{j\omega})$ by a factor L. As seen from Fig. 4.8b, the interpolator adds copies of the input spectrum in the output. These compressed copies are called *images*. There will be $L - 1$ number of images in the spectrum between 0 and 2π.

Reconstruction of Decimated Signal

Now go back to the decimation process. We had seen that removal of samples from the input signal makes the process irreversible. But we can reconstruct a signal, which is almost same as the input signal subject to some conditions. First is that there should not be aliasing as a result of decimation, i.e., the signal should be bandlimited to $|\omega| < \pi/M$. Then we have to interpolate the signal by a factor which is equal to the decimation factor, i.e., $L = M$. After interpolation, use an appropriate filter (a low-pass filter) to remove the images. The resulting signal will be same as our original signal. The process is illustrated in Fig. 4.9 with the same example we had taken for decimation and interpolation.

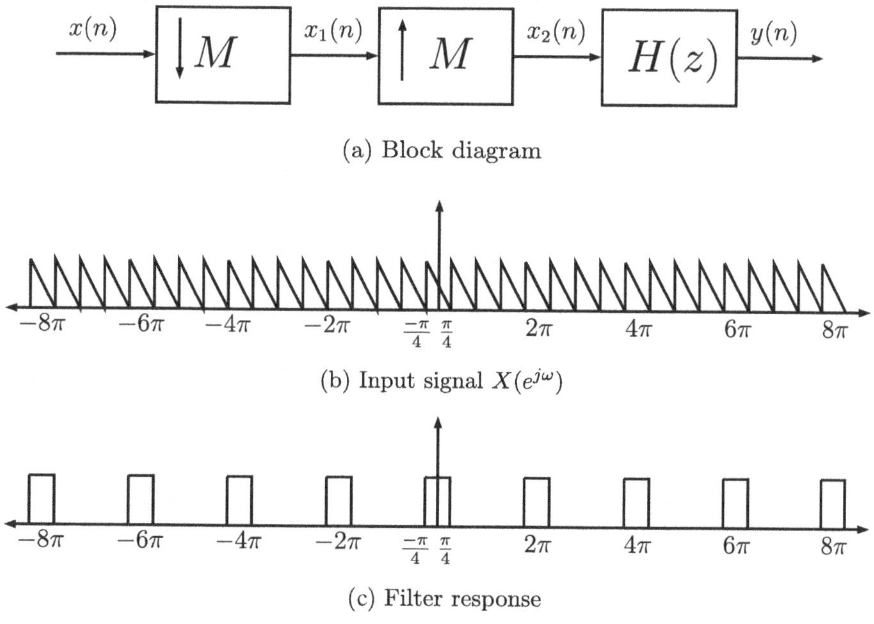

Fig. 4.9 Reconstruction of original signal from decimated signal

Quadrature Mirror Filter (QMF) Bank

We have seen that in an analysis bank each subband signal can be sampled at a lower rate than the original. Hence, we can use a decimator in each branch after filtering. Equivalently at the synthesis side, an upsampler is used followed by a filter to recover the signal. So the structure of a filter bank consists of filters, decimators, and interpolators. The typical two channel quadrature mirror filter (QMF) bank is shown in Fig. 4.10.

The QMF bank consists of two channels each having quadrature bandwidth (0 to $\frac{\pi}{2}$ and $\frac{\pi}{2}$ to π) and is symmetrical about $\frac{\pi}{2}$. The input signal is divided into two subbands by the filters $H_0[z]$ and $H_1[z]$. The filters $H_0[z]$ and $H_1[z]$ are called analysis filters. Then these signals are downsampled by a factor of two. At the synthesis side, the operations are just reverse of that in the analysis side. The signals are first upsampled by a factor of two and are then given to the filters $F_0[z]$ and $F_1[z]$ which are then combined to produce the output signal $\hat{x}[n]$. The filters $F_0[z]$ and $F_1[z]$ are called synthesis filters. The response of analysis filters $H_0[z]$ and $H_1[z]$ is shown in Fig. 4.11.

The frequency responses of two analysis filters are mirror images of each other with respect to the quadrature frequency $\frac{\pi}{2}$. Hence, the name quadrature mirror filter bank. In other words, the response $H_1[z]$ is equal to the response of $H_0[z]$ reversed and shifted by π. Mathematically, $H_1[z] = H_0[-z]$. Figure 4.11 shows that the frequency response of the filters is overlapping. The decimation of the subband signals generated by these filters surely will result in aliasing. Another option is

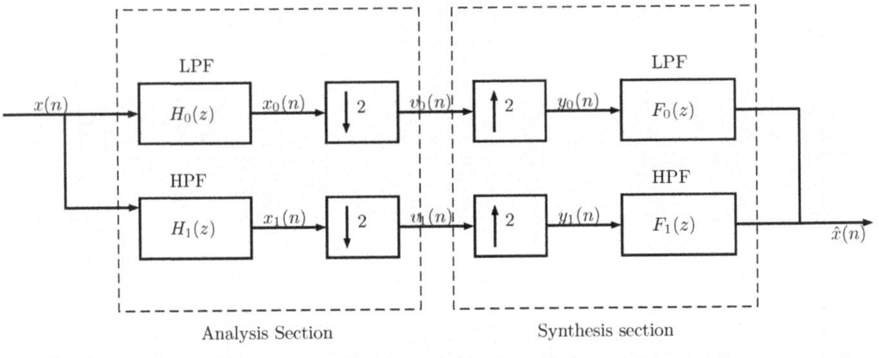

Fig. 4.10 Two channel quadrature mirror filter bank

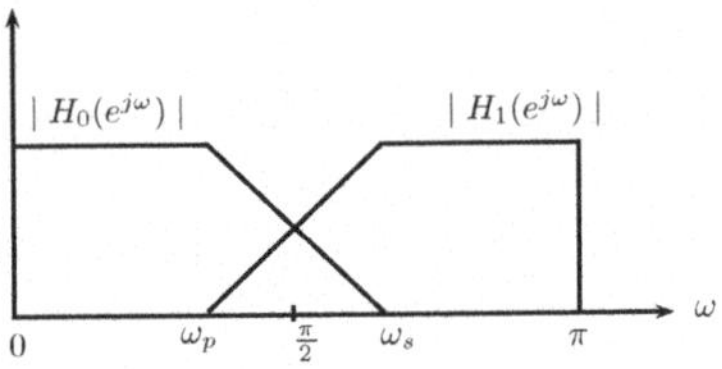

Fig. 4.11 Frequency response of analysis filters in a QMF bank

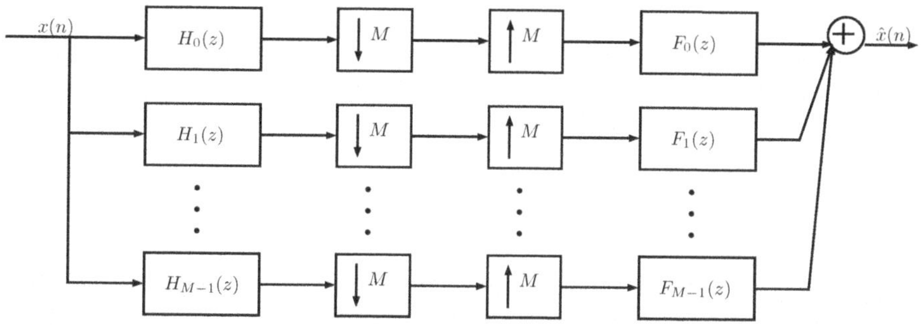

Fig. 4.12 M channel maximally decimated filter bank

to choose the filter responses as nonoverlapping by giving sharp stop bands. But this results in attenuation of signals around $\frac{\pi}{2}$. Hence, overlapping is introduced knowingly in the analysis filters. We will see in a while that even then we can avoid aliasing by proper choice of synthesis filters.

Note that the output of the filter bank is denoted by $\hat{x}[n]$ and not $x[n]$ itself. This indicates that the output is slightly different from the input as we will see in the next section.

QMF filter bank is a two channel filter bank. A filter bank can have more than two channels too. If the upsampling and the downsampling factors are equal to the number of bands, such types of filter banks are called maximally decimated or critically decimated filter banks. If M_k corresponds to the decimation factor in the kth branch, then in the case of maximally decimated filter bank, $\sum_{k=0}^{M-1} \frac{1}{M_k} = 1$. A maximally decimated filter bank with M channels is shown in Fig. 4.12.

QMF filter bank is an example of maximally decimated filter bank with $M = 2$.

Perfect Reconstruction Filter Bank

The output of QMF filter bank is different from the actual input signal because of aliasing, amplitude, and phase distortion. Perfect reconstruction means to construct the output without these distortions. For a realizable system, there will be some delay associated with it. It cannot be removed. So the output will be delayed with respect to input, i.e., there is a phase difference between the input and the output. The perfect reconstruction system allows this phase difference. Here we will discuss the conditions that are to be satisfied by the filters so as to get a perfect reconstructed signal at the output.

The subband signals from the analysis end are to be transmitted to the receiver. The receiver should reconstruct the original signal from the received signal. For effective communication, we will have to encode the signals before they are transmitted and have to decode the signals after reception. There may be slight differences between the decoded signal at the receiver end and the transmitted signal, whatever may be the efficiency of the coding scheme adopted. This error, known probably as quantization and coding error, could not be corrected. If the quantization and coding errors are neglected, three types of errors are caused by the filter bank. They are:

- Aliasing distortion
- Amplitude distortion
- Phase distortion

A system which is free from all these three distortions is known as a **perfect reconstruction (PR)** system.

Aliasing Distortion
A practical analysis filter will have a nonzero transition band, which results ultimately in aliasing. The response of a practical analysis filter is shown in Fig. 4.11. The aliasing has to be avoided to make the system PR. For that, let us find out the equation for the output signal. Referring to Fig. 4.10,

$$\begin{aligned} X_0[z] &= X[z]H_0[z] \\ X_1[z] &= X[z]H_1[z] \end{aligned} \tag{4.10}$$

By substituting $M = 2$ from Eq. 4.5,

$$\begin{aligned} V_0[z] &= \frac{1}{2}\sum_{k=0}^{1} X_0[z^{\frac{1}{2}} W_2^{-k}] \\ &= \frac{1}{2}[X[z^{\frac{1}{2}}]H_0[z^{\frac{1}{2}}] + X[-z^{\frac{1}{2}}]H_0[-z^{\frac{1}{2}}]] \end{aligned} \tag{4.11}$$

since $W_2^0 = 1$ and $W_2^{-1} = -1$.
Similarly,

$$\begin{aligned} V_1[z] &= \frac{1}{2}\sum_{k=0}^{1} X_1[z^{\frac{1}{2}} W_2^{-k}] \\ &= \frac{1}{2}[X[z^{\frac{1}{2}}]H_1[z^{\frac{1}{2}}] + X[-z^{\frac{1}{2}}]H_1[-z^{\frac{1}{2}}]] \end{aligned} \tag{4.12}$$

Neglecting any coding errors, assume the same transmitted signals reach the receiver. So

$$\begin{aligned} Y_0[z] &= \frac{1}{2}[X[z]H_0[z] + X[-z]H_0[-z]] \\ Y_1[z] &= \frac{1}{2}[X[z]H_1[z] + X[-z]H_1[-z]] \end{aligned} \tag{4.13}$$

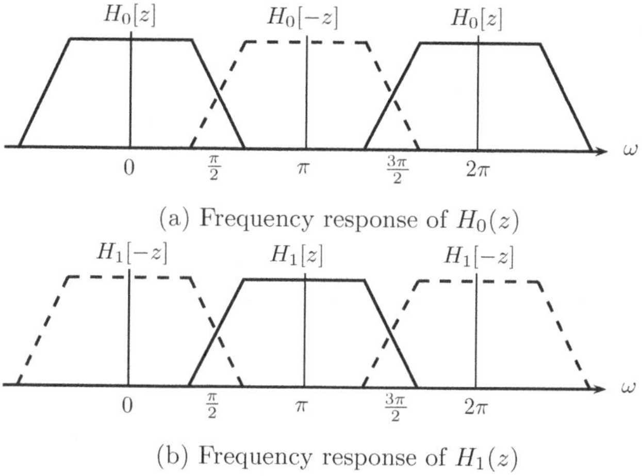

Fig. 4.13 Figure showing analysis and synthesis filter designs

Combining the above two signals

$$\begin{aligned} Y[z] &= F_0[z]\left\{\frac{1}{2}[X[z]H_0[z] + X[-z]H_0[-z]]\right\} \\ &+ F_1[z]\left\{\frac{1}{2}[[X[z]H_1[z] + X[-z]H_1[-z]]\right\} \\ &= \frac{1}{2}\left\{F_0[z]H_0[z] + F_1[z]H_1[z]\right\}X[z] \\ &+ \frac{1}{2}\left\{F_0[z]H_0[-z] + F_1[z]H_1[-z]\right\}X[-z] \end{aligned} \tag{4.14}$$

The second term containing $X[-z]$ originated by decimation and interpolation is responsible for aliasing. So, to remove aliasing, we have to make the second term zero. By having a look at the equation, clearly the second term is zero when,

$$F_0[z] = H_1[-z] \quad \text{and} \quad F_1[z] = -H_0[-z] \tag{4.15}$$

Thus, in QMF filter bank, by allowing aliasing in analysis filters, we can reconstruct our original signal by proper choice of synthesis filters.

Now we will discuss the physical meaning of alias cancelation. Aliasing occurs due to the overlapping in the spectrum of analysis filters. We are designing synthesis filters in such a way that the effect of aliasing is removed. Figure 4.13 explains the concept better.

$H_0[-z]$ is the response of H_0 shifted by π. Similarly, $H_1[-z]$ is the shifted version of $H_1[z]$ by π. Note that the spectrum is repeating. So by taking $F_1[z] = -H_0[-z]$, the aliasing can be removed. Thus we can conclude that by permitting some amount of aliasing in the analysis side, and by the proper design of synthesis filters, aliasing can be canceled.

We can write Eq. 4.13 as

$$Y[z] = X[z]T[z] + X[-z]A[z] \tag{4.16}$$

$A[z]$ term should be equal to zero to cancel aliasing. So $Y[z] = X[z]T[z]$. $T[z]$ is known as the distortion transfer function.

$$\begin{aligned} T[z] &= \frac{1}{2}[[F_0[z]H_0[z] + F_1[z]H_1[z]]] \\ &= \frac{1}{2}[[H_0[z]H_1[-z] - H_1[z]H_0[-z]]] \end{aligned} \tag{4.17}$$

Amplitude and Phase Distortion

We have to make the term $T[z]$ free of amplitude and phase distortion. If all the samples of the input signal are multiplied by the same factor, then there will be no amplitude distortion. The best choice for this purpose is an all-pass transfer function. If the analysis filter $H_0[z]$ is made a linear phase FIR filter, then $T[z]$ will also have linear phase. For that, the impulse response $h_0[n]$ should be symmetric, i.e., $h_0[n] = h_0[N-1-n]$. In this case, there will be no phase distortion. Combining these two conditions, the output should be a scaled and shifted version of the input signal. So,

$$T[z] = cz^{-n_0} \quad \text{or} \quad y[n] = cx[n-n_0] \tag{4.18}$$

With the selection of $h[n]$ to be symmetric,

$$H_0(e^{j\omega}) = e^{-j\omega(\frac{N-1}{2})}H_r(\omega) \tag{4.19}$$

where $H_r(\omega)$ is the amplitude function. So we can write

$$T(e^{j\omega}) = \frac{1}{2}e^{-j\omega(N-1)}[|H_0(e^{j\omega})|^2 - (-1)^{N-1}|H_1(e^{j(\pi-\omega)})|^2] \tag{4.20}$$

The above equation becomes 0 when N is even at $\omega = \pi/2$ which causes severe amplitude distortion. Hence N should be odd. So

$$T(e^{j\omega}) = \frac{1}{2}e^{-j\omega(N-1)}[|H_0(e^{j\omega})|^2 + |H_1(e^{j(\pi-\omega)})|^2] \tag{4.21}$$

For a PR system

$$|T(e^{j\omega})| = |H_0(e^{j\omega})|^2 + |H_0(e^{j(\pi-\omega)})|^2 = c \tag{4.22}$$

A system which is free from aliasing, amplitude, and phase distortion can be used for perfect reconstruction.

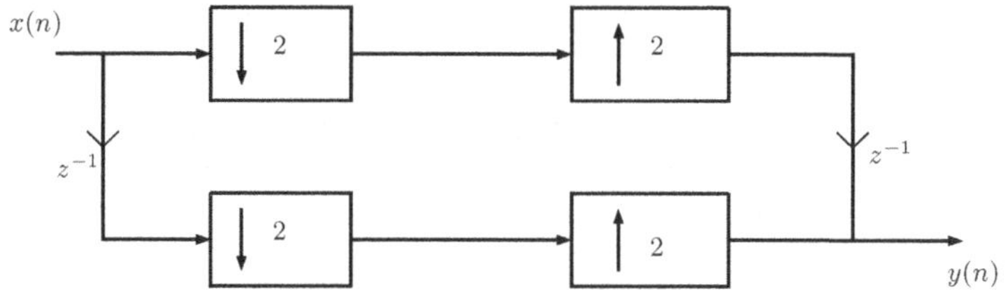

Fig. 4.14 Example of a PR QMF bank

From the mirror image structure,

$$H_1[z] = H_0[-z] \tag{4.23}$$

This means that if $H_0[z]$ is a good low-pass filter, then $H1_[z]$ will be a good high-pass filter. So by using Eq. 4.15,

$$F_0[z] = H_0[z] \quad \text{and} \quad F_1[z] = -H_0[-z] \tag{4.24}$$

Hence, the knowledge of a single filter $H_0[z]$ is enough to design the whole filter bank.

$$\begin{aligned} T[z] &= \frac{1}{2}\,[[H_0[z]H_1[-z] - H_1[z]H_0[-z]]] \\ &= \frac{1}{2}\,[[H_0[z]H_0[z] - H_0[-z]H_0[-z]]] \\ &= \frac{1}{2}\left[[H_0^2[z] - H_0^2[-z]]\right] \end{aligned} \tag{4.25}$$

Example of a PR System
Consider a QMF bank with $H_0(z) = 1$, $H_1(z) = z^{-1}$, $F_0(z) = z^{-1}$ and $F_1(z) = 1$. Substituting the above values in Eq. 4.14, we get $Y(z) = z^{-1}X(z)$ and $T(z) = z^{-1}$. So the system is an example of a PR system and is demonstrated in Fig. 4.14.

Polyphase Representation

It will be useful if we can represent analysis and synthesis bank in terms of polyphase components. Let

$$H_0[z] = E_0[z^2] + z^{-1}E_1[z^2] \tag{4.26}$$

Substituting $H_1[z] = H_0[-z]$ in the above equation, we get

$$H_1[z] = E_0[z^2] - z^{-1}E_1[z^2] \tag{4.27}$$

Thus we can write

$$\begin{bmatrix} H_0[z] \\ H_1[z] \end{bmatrix} = \begin{bmatrix} 1 & 1 \\ 1 & -1 \end{bmatrix} \begin{bmatrix} E_0[z^2] \\ z^{-1}E_1[z^2] \end{bmatrix} \tag{4.28}$$

Using Eq. 4.24, the expression for synthesis filters can be written as

$$\begin{aligned} F_0[z] &= H_0[z] = E_0[z^2] + z^{-1}E_1[z^2] \\ F_1[z] &= -H_1[z] = -E_0[z^2] + z^{-1}E_1[z^2] \end{aligned} \tag{4.29}$$

They can be represented in terms of polyphase components as

$$\begin{bmatrix} F_0[z] \; F_1[z] \end{bmatrix} = \begin{bmatrix} z^{-1}E_1[z^2] \; E_0[z^2] \end{bmatrix} \begin{bmatrix} 1 & 1 \\ 1 & -1 \end{bmatrix} \tag{4.30}$$

Using the polyphase equation, the analysis and synthesis bank can be redrawn as shown in Fig. 4.15. The combined QMF filter is shown in Fig. 4.16. The complete

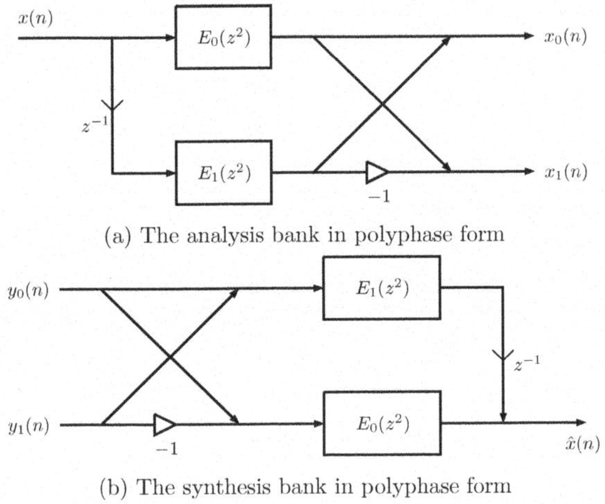

Fig. 4.15 Polyphase representation of analysis and synthesis bank

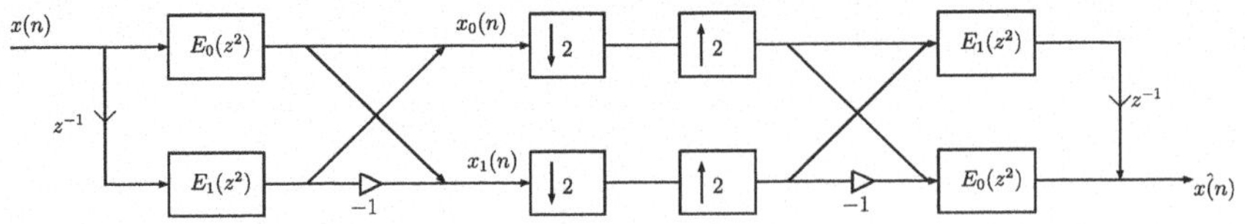

Fig. 4.16 Complete QMF bank polyphase form

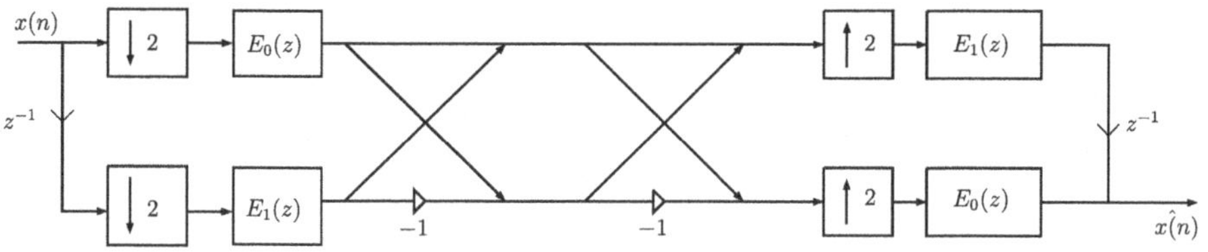

Fig. 4.17 QMF bank modified using noble identities

structure can be modified using noble identities, and the resultant structure is shown in Fig. 4.17. Substituting Eq. 4.25 in Eq. 4.26, we get

$$T[z] = 2z^{-1}E_0(z^2)E_1(z^2) \tag{4.31}$$

From the above equation, it is clear that if $H_0(z)$ is FIR, then $E_0(z)$, $E_1(z)$, and $T(z)$ all should be FIR. Also in order to avoid amplitude distortion, $E_0(z)$ and $E_1(z)$ should be a delay, i.e., $E_0(z) = c_0z^{-n_0}$ and $E_1(z) = c_1z^{-n_1}$.

4.2 M Channel Filter Bank

Till now we discussed the two channel QMF filter bank with $H_1(z) = H_0(-z)$. Now we will look into a more general case, to fully understand the concept of alias cancelation and perfect reconstruction. For that, consider a general M channel maximally decimated filter bank as shown in Fig.4.18.

The input signal $x(n)$ is divided into M subband signals denoted by $x_k(n)$ by the corresponding analysis filters $H_k(z)$. Each subband signal is then decimated by a factor of M to obtain $v_k(n)$. At the synthesis side, the subband signals are first passed through M-fold expanders and recombined using synthesis filters $F_k(z)$ to produce $\hat{x}(n)$. The analysis bank is an one-input M-output system with transfer matrix $h(z)$, and the synthesis bank is a M-input one-output system with transfer matrix $f^T(z)$.

Each subband signal $X_k(z)$ is given by

$$X_k(z) = H_k(z)X(z) \tag{4.32}$$

and the decimated signal

$$V_k(z) = \frac{1}{M}\sum_{l=0}^{M-1} H_k(z^{\frac{1}{M}}W^l)X(z^{\frac{1}{M}}W^l) \tag{4.33}$$

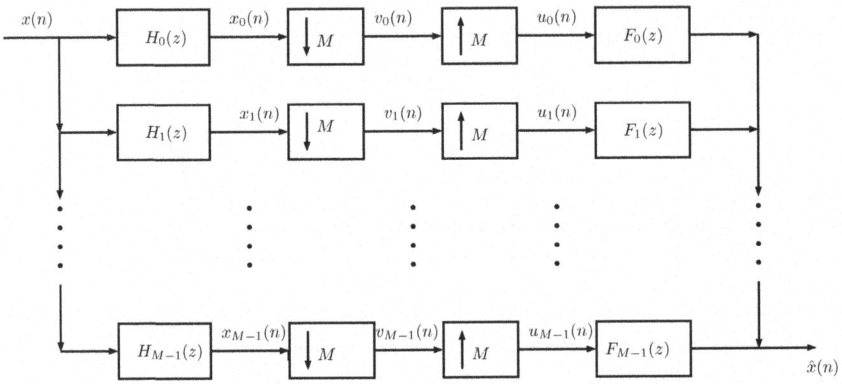

Fig. 4.18 General M channel QMF bank

The signals at the synthesis side are given by

$$U_k(z) = V_k(z^M) = \frac{1}{M}\sum_{l=0}^{M-1} H_k(zW^l)X(zW^l)$$
$$\hat{X}(z) = \sum_{k=0}^{M-1} F_k(z)U_k(z) = \frac{1}{M}\sum_{l=0}^{M-1} X(zW^l)\sum_{k=0}^{M-1} H_k(zW^l)F_k(z) \tag{4.34}$$

The reconstructed signal can be easily represented as

$$\hat{X}(z) = \sum_{l=0}^{M-1} A_l(z)X(zW^l) \tag{4.35}$$

where

$$A_l(z) = \frac{1}{M}\sum_{k=0}^{M-1} H_k(zW^l)F_k(z) \tag{4.36}$$

Here also as in the case of two channel QMF bank, the reconstructed signal differs from the original signal due to aliasing, amplitude distortion, and phase distortion. From the above equations, it is clear that $X(zW^l)$ is the lth aliasing term, and its gain is given by $A_l(z)$. So aliasing can be avoided by making sure that

$$A_l(z) = 0,\ 1 \le l \le M-1 \tag{4.37}$$

When aliaing is canceled, the distortion function can be written as

$$T(z) \approx A_0(z) = \frac{1}{M}\sum_{k=0}^{M-1} H_k(z)F_k(z) \tag{4.38}$$

In order to avoid amplitude distortion, $T(z)$ should be all pass, and to avoid phase distortion, $T(z)$ should have linear phase.

Just as in the case of two channel QMF, if $H_k(z)$ and $F_k(z)$ are such that aliasing is completely canceled and if $T(z)$ is a pure delay, then the system is called **perfect reconstruction (PR)** system and will be free from aliasing, amplitude, and phase distortion.

Polyphase Representation

Any transfer function $H_k(z)$ can be represented in type 1 polyphase form as

$$H_k(z) = \sum_{l=0}^{M-1} z^{-l} E_{kl}(z^M) \tag{4.39}$$

The same can be represented in matrix form as

$$\begin{bmatrix} H_0[z] \\ \vdots \\ H_{M-1}[z] \end{bmatrix} = \begin{bmatrix} E_{00}[z^M] & E_{01}[z^M] & \cdots & E_{0,M-1}[z^M] \\ \vdots & \vdots & \ddots & \vdots \\ E_{M-1,0}[z^M] & E_{M-1,1}[z^M] & \cdots & E_{M-1,M-1}[z^M] \end{bmatrix} \begin{bmatrix} 1 \\ z^{-1} \\ \vdots \\ z^{-(M-1)} \end{bmatrix} \tag{4.40}$$

i.e.,

$$h[z] = E[z^M]e[z] \tag{4.41}$$

where $h[z]$ is the vector representing the analysis bank given by $h[z] = \begin{bmatrix} H_0[z] \\ H_1[z] \\ \vdots \\ H_{M-1}[z] \end{bmatrix}$, $e[z]$ is the vector representing the delay chain given by $e[z] =$

$\begin{bmatrix} 1 \\ z^{-1} \\ \vdots \\ z^{-(M-1)} \end{bmatrix}$, and $E[z]$ is the MXM type 1 polyphase matrix for the analysis bank given by

$$E[z] = \begin{bmatrix} E_{00}[z] & E_{01}[z] & \cdots & E_{0,M-1}[z] \\ E_{10}[z] & E_{11}[z] & \cdots & E_{1,M-1}[z] \\ \vdots & \vdots & \ddots & \vdots \\ E_{M-1,0}[z] & E_{M-1,1}[z] & \cdots & E_{M-1,M-1}[z] \end{bmatrix} \tag{4.42}$$

The synthesis filers can also be represented in polyphase form in a similar manner.

$$F_k[z] = \sum_{l=0}^{M-1} z^{-(M-1-l)} R_{lk}[z^M] \tag{4.43}$$

In matrix form,

$$\begin{aligned} &\left[F_0[z] \cdots F_{M-1}[z]\right] \\ &= \left[z^{-(M-1)}\ z^{-(M-2)}\ \ldots\ 1\right] \begin{bmatrix} R_{00}[z^M] & R_{01}[z^M] & \cdots & R_{0,M-1}[z^M] \\ R_{10}[z^M] & R_{11}[z^M] & \cdots & R_{1,M-1}[z^M] \\ \vdots & \vdots & \ddots & \vdots \\ R_{M-1,0}[z^M] & R_{M-1,1}[z^M] & \cdots & R_{M-1,M-1}[z^M] \end{bmatrix} \end{aligned} \tag{4.44}$$

i.e.,

$$f^T[z] = z^{-(M-1)}\tilde{e}[z]R[z^M] \tag{4.45}$$

where $f[z]$ is the vector representing the synthesis bank given by $f[z] = \begin{bmatrix} F_0[z] \\ F_1[z] \\ \vdots \\ F_{M-1}[z] \end{bmatrix}$, $e[z]$ is the vector representing the delay chain given by $e[z] = \begin{bmatrix} 1 \\ z^{-1} \\ \vdots \\ z^{-(M-1)} \end{bmatrix}$, and $R[z]$ is the MXM type 2 polyphase matrix for the synthesis bank given by

$$R[z] = \begin{bmatrix} R_{00}[z] & R_{01}[z] & \cdots & R_{0,M-1}[z] \\ R_{10}[z] & R_{11}[z] & \cdots & R_{1,M-1}[z] \\ \vdots & \vdots & \ddots & \vdots \\ R_{M-1,0}[z] & R_{M-1,1}[z] & \cdots & R_{M-1,M-1}[z] \end{bmatrix} \tag{4.46}$$

Figure 4.19a shows the polyphase representation of an M channel filter bank. The same can be reduced using noble identities to obtain the figure in Fig. 4.19b. The two matrices $E[z]$ and $R[z]$ are combined into a single matrix $P[z]$ which is defined by

$$P[z] = R[z]E[z] \tag{4.47}$$

The system can be redrawn as in Fig. 4.19c using Eq. 4.47.

Perfect Reconstruction System

An example of a general M channel perfect reconstruction system can be like

$$H_k(z) = z^{-k},\ F_k(z) = z^{-(M-1-k)},\ 0 \leq k \leq M-1 \tag{4.48}$$

Substituting in Eq. 4.14, we get

$$\begin{aligned} \hat{X}(z) &= z^{-(M-1)}X(z) \\ \text{i.e.,}\quad \hat{x}(n) &= x(n-M+1) \\ T(z) &= z^{-(M-1)} \end{aligned} \tag{4.49}$$

A general PR M channel filter bank is shown in Fig. 4.20.

In terms of polyphase components $E(z)$ and $R(z)$, the PR condition can be stated as

$$R[z]E[z] = I \tag{4.50}$$

More generally, we can say that for a PR system,

$$R[z]E[z] = cz^{-m_0}I \tag{4.51}$$

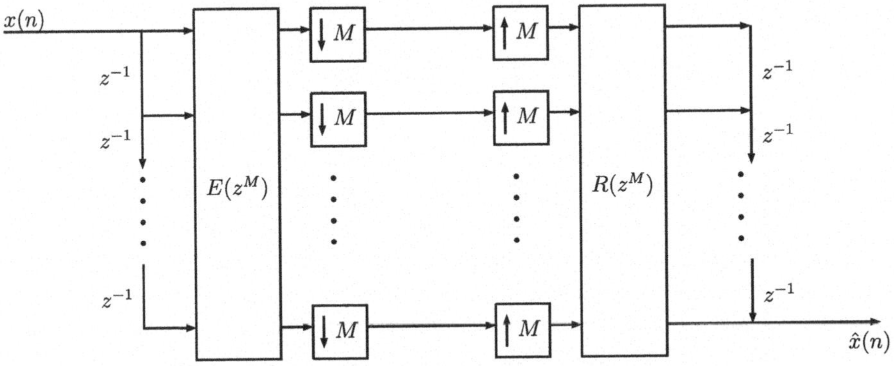

(a) Polyphase representation of an M channel filter bank

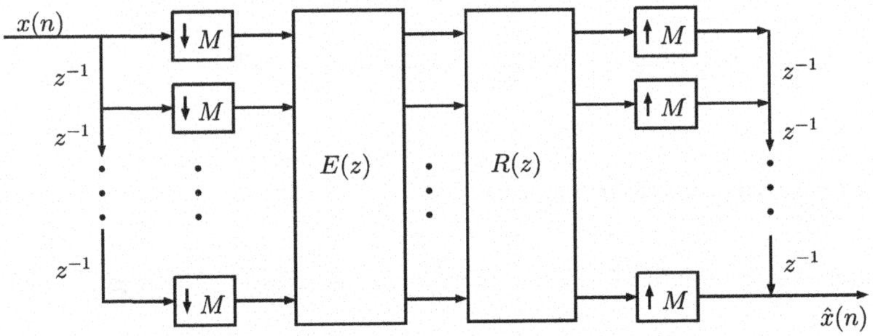

(b) Modified diagram using noble identities

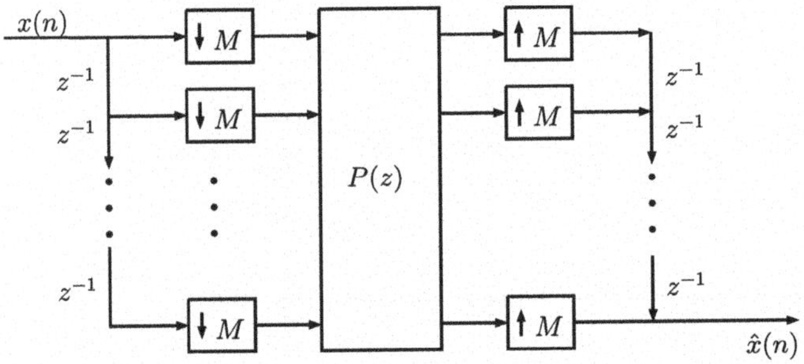

(c) Further modified diagram using Eqn.4.47

Fig. 4.19 Polyphase representation of an M channel filter bank

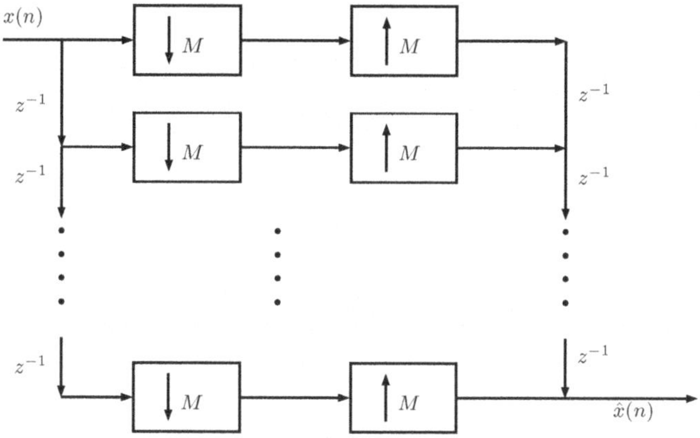

Fig. 4.20 M channel perfect reconstruction filter bank

which can again be modified into

$$R[z]E[z] = cz^{-m_0} \begin{bmatrix} 0 & I_{M-r} \\ z^{-1} I_r & 0 \end{bmatrix} \tag{4.52}$$

for some integer r with $0 \leq r \leq M - 1$, some integer m_0, and some constant $c \neq 0$. It can be derived from this equation that

$$|R[z]||E[z]| = c_0 z^{-k_0} \tag{4.53}$$

for some $c_0 \neq 0$ and some integer k_0. Every PR system irrespective of FIR or IIR must satisfy the above condition.

4.3 Tree Structured Filter Banks

In many applications, another structure of filter banks called tree structured filter bank is used. Consider a structure as shown in Fig. 4.21a. In this structure, first the signal is split into two subband signals, and each subband signal is again divided into two after decimation. At the synthesis side, the signals are recombined

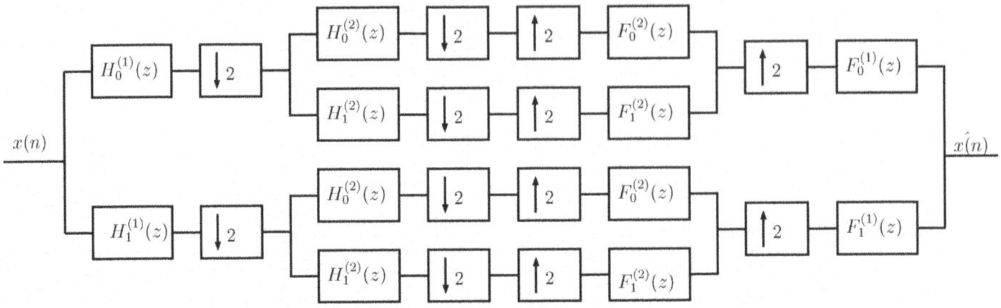

(a) Two level maximally decimated tree stuctured filter bank

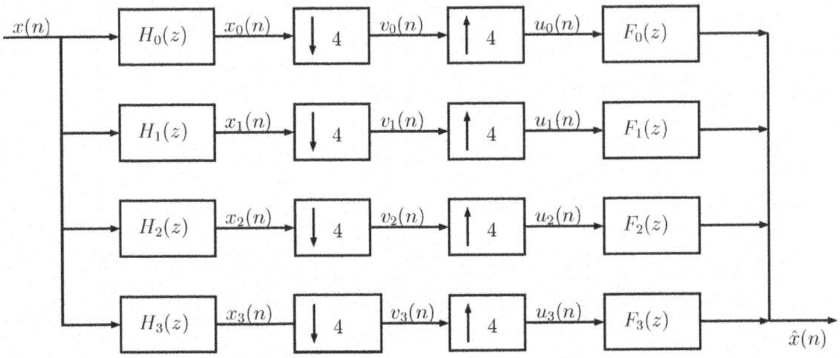

(b) Equivalent non-tree form

Fig. 4.21 Tree structured filter bank

two at a time. This system is a maximally decimated tree structured filter bank. This complete system is equivalent to the non-tree structure form as shown in Fig. 4.21b.

If the two channel QMF bank with the filters shown in figure is alias free, then the complete system is also alias free. Similarly, if the two channel QMF is PR, then the complete system is also PR.

We can have tree structure with more than two levels of splitting. Also flexibility can be permitted in the number of divisions in each subband. We can split a subband into two, then can split one subband into three, another into four, and so on. One such structure which is more popular in multiresolution analysis in image processing is shown in Fig. 4.22.

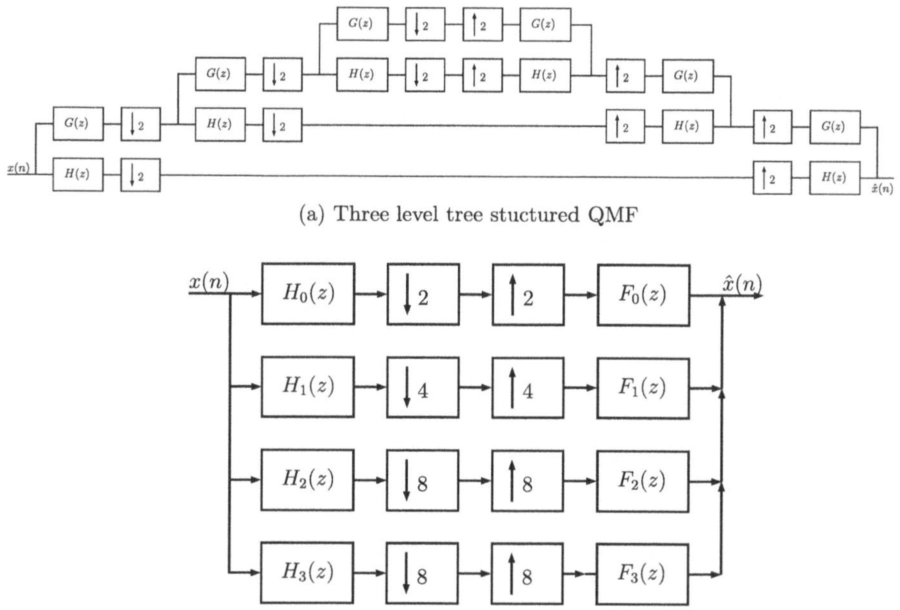

(a) Three level tree stuctured QMF

(b) Equivalent non-tree form

Fig. 4.22 Polyphase representation of an M channel filter bank

4.4 DWT Realization Using Filter Bank

Now we can discuss the realization of discrete wavelet transform (DWT) using filter banks. As mentioned in the beginning of this chapter, the filter bank implementation of discrete wavelet transform is very useful in many applications as it makes the analysis and implementation easier.

Discrete wavelet transform is different from the Fourier transform because DWT can provide both time and frequency resolutions simultaneously. Actually what we are doing in discrete Fourier transform? We are representing a discrete signal in terms of continuous sinusoids. DWT is just the opposite of that, i.e., we are representing a continuous time signal in terms of discontinuous signals called wavelets. How can we represent a continuous signal in terms of discontinuous functions? Using piecewise constant functions. So, what is this piecewise constant function? Assume that we have an audio signal which is essentially continuous in time. Now divide the time axis into equal intervals of T as shown in Fig. 4.23. For each interval, we can assign a constant value to represent the audio signal in that interval. The suitable constant value is the average value of the signal in that interval. This is known as the piecewise constant representation.

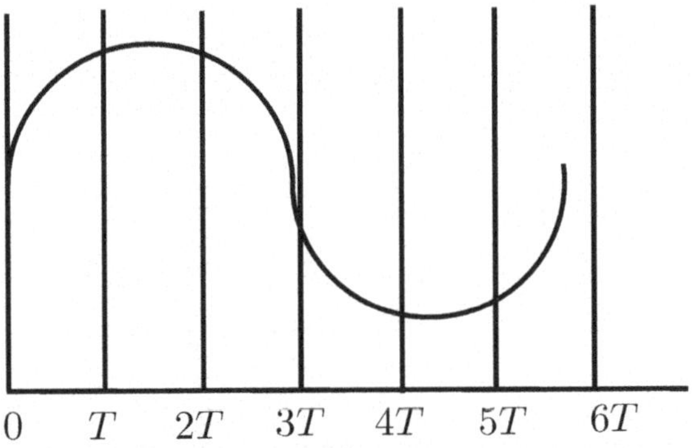

Fig. 4.23 An audio signal

If the value of T is equal to one, or if we are considering piecewise constant approximation on unit intervals, we get the space V_0 (mentioned in the second chapter). Similarly, if the value of T is equal to $\frac{1}{2}$ or if we are considering piecewise constant approximation on half intervals, we get the space V_1. Generally, V_n is the space of piecewise constant approximations on 2^{-n} intervals. The word space is used because the linear combination of functions in this set will get back to a function enclosed in the same set. By using intervals of half size, the approximations to the original signal will be more accurate. In other words, more information can be captured by half intervals than unit intervals. Hence, V_1 is larger than V_0. A function which is piecewise constant in unit intervals is automatically constant in half intervals. Combining these two conditions, we can write

$$\cdots V_{-2} \subset V_{-1} \subset V_0 \subset V_1 \subset V_2 \cdots \tag{4.54}$$

The additional information obtained by going from V_0 to V_1 is represented by W_0. So the space $V_j s$ are called approximation spaces and $W_j s$ are called detail spaces. The relation between $W_j s$ is given by

$$\cdots W_{-2} \perp W_{-1} \perp W_0 \perp W_1 \perp W_2 \cdots \tag{4.55}$$

Haar Filter Bank

To derive the filter bank structure, it is easier to start with the Haar and then to generalize. Haar scaling and wavelet functions are given in Fig. 4.24.

Haar scaling function $\phi(t)$ is nonzero only in the interval [0, 1]. Its dilates and translates can span the whole space V_0. Similarly, V_1 is spanned by the dilates and translates of $\phi(2t)$. Consider the following example.

Fig. 4.24 Haar scaling and wavelet functions

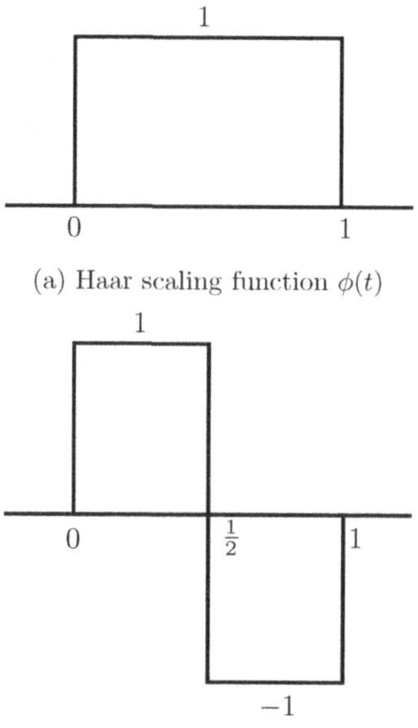

(a) Haar scaling function $\phi(t)$

(b) Haar wavelet function $\psi(t)$

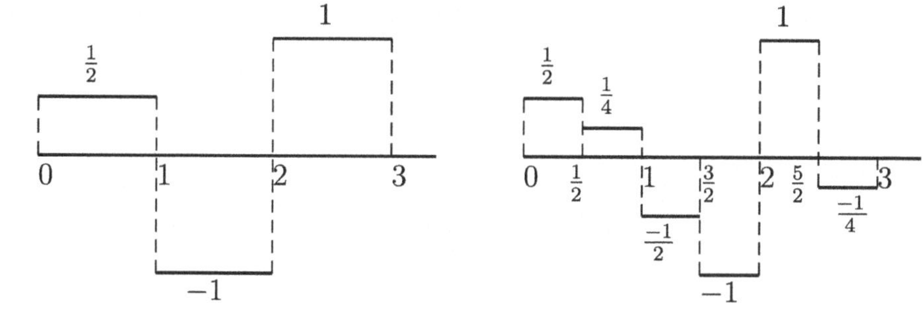

(a) Piecewise constant approximation on unit intervals

(b) Piecewise constant approximation on half intervals

Fig. 4.25 Example of piecewise constant functions in V_0 and V_1

Figure 4.25a represents a piecewise constant function in V_0. It can be represented in terms of scaling function $\phi(t)$ as

$$\frac{1}{2}\phi(t) + (-1)\phi(t-1) + (1)\phi(t-2) \tag{4.56}$$

Hence, we say that $\phi(t)$ spans V_0. Similarly, Fig. 4.25b represents a piecewise constant function in V_1. So it can be represented as translates and dilates of $\phi(2t)$. So the signal can be written as

$$\frac{1}{2}\phi(2t)+\frac{1}{4}\phi(2t-1)-\frac{1}{2}\phi(2t-2)-1\phi(2t-3)+\phi(2t-4)-\frac{1}{4}\phi(2t-5) \tag{4.57}$$

Any function in V_1 can be represented in terms of $\phi(2t)$. Hence, $\phi(2t)$ spans V_1. Extending up to m, V_m is spanned by translates and dilates of $\phi(2^m t)$.

$$V_m = span\{\phi(2^m t - n)\} \tag{4.58}$$

The scaling function $\phi(t)$ forms an orthogonal basis. That is, the inner product of scaling function with its translates is zero.

$$< \phi(t-n), \phi(t-m) > \quad = \quad \begin{cases} 1, & \text{if} \quad \text{n=m} \\ 0, & \text{if} \quad \text{n} \neq \text{m} \end{cases} \tag{4.59}$$

The wavelet function $\psi(t)$ and its translates span the space W_0 and so on. So the wavelet function helps to capture additional information when going from one resolution to other, whereas the scaling function captures signals at a particular resolution. Also note that $\phi(t)$ is orthogonal to $\psi(t)$. The space V_1 is said to be the orthogonal sum of the spaces V_0 and W_0 if there is a unique way to take a vector in V_1 and decompose it as a sum of a vector in V_0 and a vector in W_0 with the vectors in V_0 and W_0 being orthogonal.

Now consider a function $a[n]$ in V_1 as shown in Fig. 4.26.

Remember that in the space V_1, the intervals are half. So from the figure, we can write

$$\begin{aligned} a[2n] &= c_1 \\ a[2n+1] &= c_2 \end{aligned} \tag{4.60}$$

Any function in V_1 can be represented as the sum of functions in V_0 and W_0. Let the function in V_0 be a_{V_0} and in W_0 be a_{W_0}. They are as shown below (Fig. 4.27).

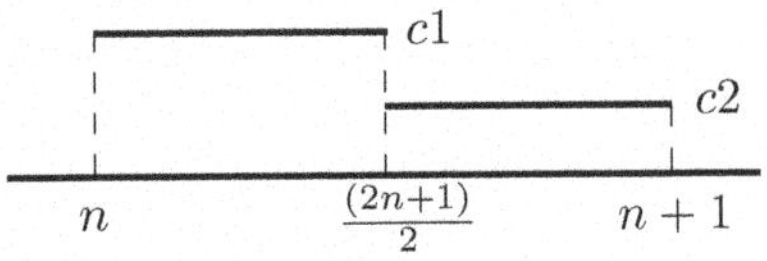

Fig. 4.26 Example function in V_1

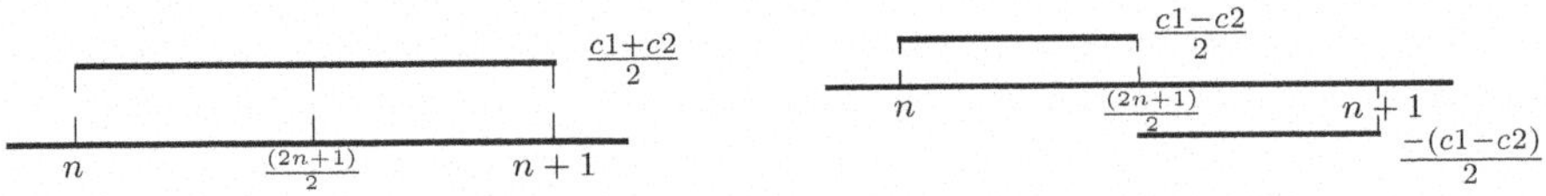

(a) Projection of $a(n)$ on V_0

(b) Projection of $a(n)$ on W_0

Fig. 4.27 Function in V_1 decomposed into functions in V_0 and W_0

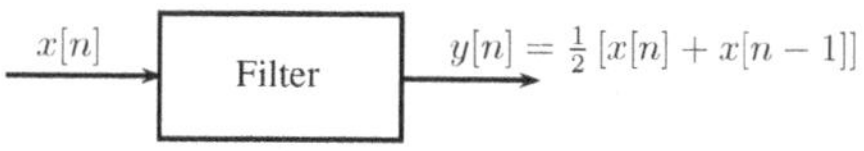

Fig. 4.28 A filter with input x(n)

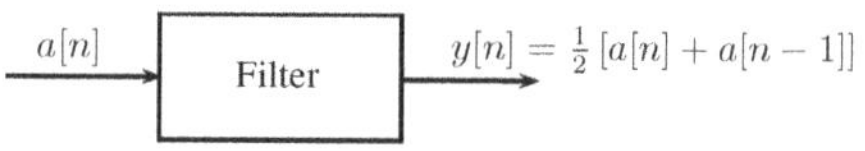

Fig. 4.29 A filter with input a(n)

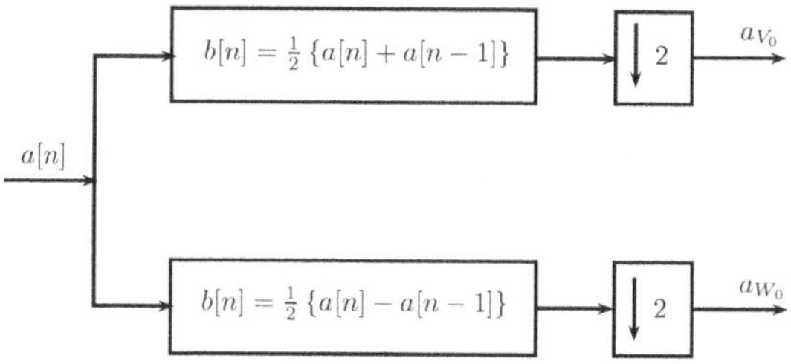

Fig. 4.30 Modified filter diagram to get $a_{V_0}(n)$ and $a_{W_0}(n)$

From the figures, it is clear that

$$\begin{aligned} a_{V_0}[n] &= \frac{c_1 + c_2}{2} \\ &= \frac{a[2n] + a[2n+1]}{2} \\ a_{W_0}[n] &= \frac{c_1 - c_2}{2} \\ &= \frac{a[2n] - a[2n+1]}{2} \end{aligned} \tag{4.61}$$

Assume that we have a filter that produces an output $y[n]$ for an input $x[n]$ (as shown in Fig. 4.28) where $y[n]$ is given by

$$y[n] = \frac{1}{2}\{x[n] + x[n+1]\} \tag{4.62}$$

Now give the input to the filter as our $a[n]$. So what will be the output? Clearly, it will be $\frac{1}{2}\{a[n] + a[n+1]\}$ (Fig. 4.29).

Compare the output equation with the equation of $a_{V_0}(n)$ given in Eq. 4.61. What is the difference between Eq. 4.32 and the output of this filter? $a[n]$ should be replaced with $a[2n]$. Which system can provide an output $a[2n]$ for an input $a[n]$? No doubt, it is a decimator with a decimation factor of 2!. By connecting a decimator at the output of the filter, we get the desired a_{V_0} and a_{W_0}. So now our structure is shown in Fig. 4.30.

A filter is always represented by its transfer function. Here $b[n] = \frac{1}{2}\{a[n] + a[n+1]\}$. Take Z transform on both sides. We get

$$B[z] = \frac{1}{2}\left\{A[z] + z^{-1}A[z]\right\} = \frac{1}{2}\left\{A[z][1 + z^{-1}]\right\} \tag{4.63}$$

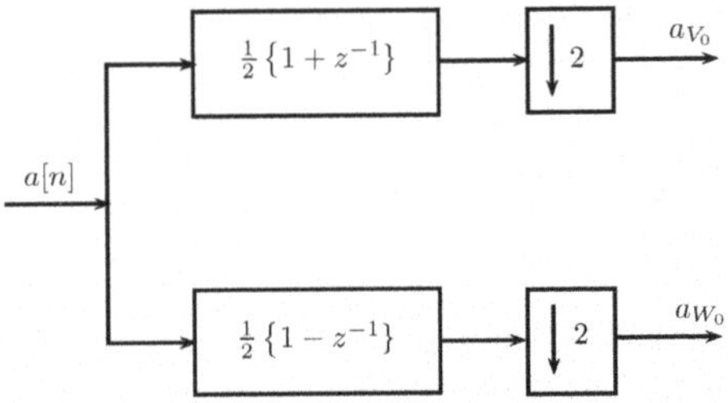

Fig. 4.31 Complete analysis filter bank structure

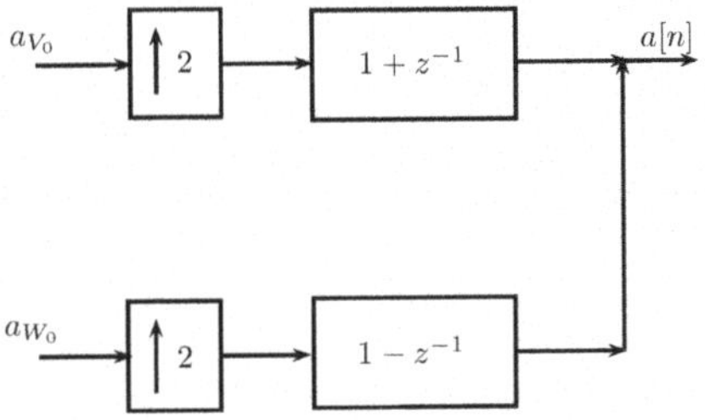

Fig. 4.32 Complete synthesis filter bank structure

So the transfer function is

$$\frac{B[z]}{A[z]} = \frac{1}{2}\left\{1 + z^{-1}\right\} \tag{4.64}$$

Similarly in the lower branch

$$\frac{B[z]}{A[z]} = \frac{1}{2}\left\{1 - z^{-1}\right\} \tag{4.65}$$

Now the complete analysis structure is shown in Fig. 4.31.

Now we can simply derive the synthesis filter bank structure. As we had decimated by a factor of 2 in the analysis side, the length of the output signal in each branch will be half that of the original signal. To compensate that, we should upsample the signal at the synthesis side first. After upsampling, the signal in the first branch will be $\{\frac{c_1+c_2}{2}, 0\}$, and that in the second branch will be $\{\frac{c_1-c_2}{2}, 0\}$. Now on addition of the signals in two branches, we get $\{c_1, 0\}$, and on subtraction, we get $\{c_2, 0\}$. The first sample of the output should be the first sample from the top branch, and the second sample of the output should be the first sample from the bottom branch. So we have to select alternatively from top and bottom branches. This can be done by introducing a delay. So on top branch, the transfer function should be $[1 + z^{-1}]$, and in bottom branch, it should be $[1 - z^{-1}]$. Thus we had the synthesis filter bank structure as shown in Fig. 4.32.

The complete filter bank structure is shown in Fig. 4.33.

It is very simple to prove that $\frac{1}{2}\left\{1 + z^{-1}\right\}$ is a low-pass filter and $\frac{1}{2}\left\{1 - z^{-1}\right\}$ is a high-pass filter as we know their frequency response.

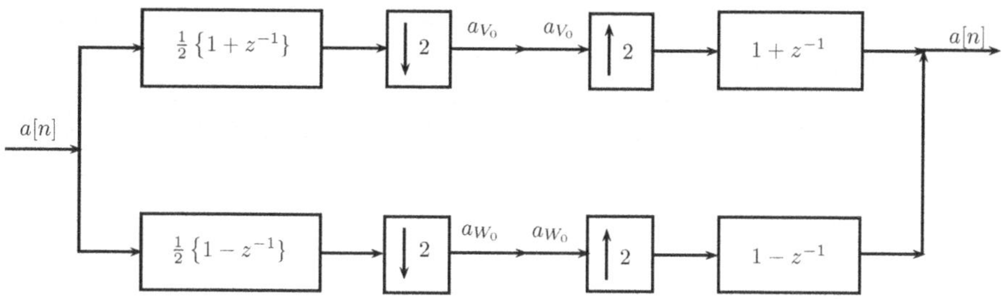

Fig. 4.33 Complete filter bank structure for Haar

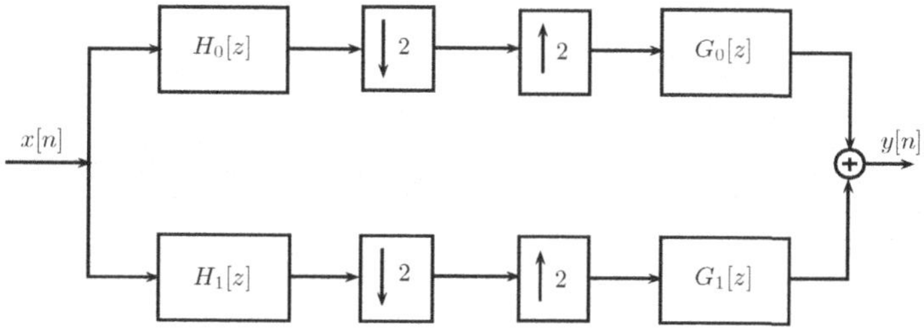

Fig. 4.34 Filter bank realization of DWT

General Filter Bank

Now we can draw the general structure of the filter bank realization of DWT as given in Fig. 4.34.

Next we have to derive the refinement relations. We know $\phi(t) \in V_0$ and $V_0 \subset V_1$. So $\phi(t)$ should be expressible in terms of $\phi(2t-n)$ which is the basis for V_1, i.e., $\phi(t)$ can be expressed in terms of its own translations and dilations. From the Haar scaling function figure shown in Fig. 4.24a, it is very easy to write

$$\phi(t) = \phi(2t) + \phi(2t-1) \tag{4.66}$$

The coefficients $\{1, 1\}$ are the coefficients of impulse response $H_0[z]$. Similarly, $\psi(t) \in V_0$ and $V_0 \subset V_1$. So $\psi(t)$ should be expressible in terms of $\phi(2t-n)$. From Fig. 4.24b showing Haar wavelet function

$$\psi(t) = \phi(2t) - \phi(2t-1) \tag{4.67}$$

Here also the coefficients $\{1, -1\}$ are the coefficients of impulse response $H_1[z]$. Generalizing this idea,

$$\begin{aligned} \phi(t) &= \sum_n h[n]\phi(2t-n) \\ \psi(t) &= \sum_n g[n]\phi(2t-n) \end{aligned} \tag{4.68}$$

where $h[n]$ and $g[n]$ are the coefficients of impulse response of the low-pass and high-pass filters, respectively. These relations are known as refinement relations. Also to make the filter bank a PR one, as we had seen earlier, $H_0[z] = H_1[-z]$. Generally, we can say $H_1[-z]$ should be related to $H_0[z]$. So we may choose $H_1[z] = z^{-D} H_0[-z-1]$. In the Haar case, $H_1[z] = z^{-1} H_0[-z-1]$. This implies that the coefficients $g[n]$ are related to $h[n]$, and hence, in order to design the whole filter bank structure, we need to know only the low-pass filter coefficients $h[n]$.

4.5 Solved Problems

Problem A sinusoidal signal $x[n]$ is shown in Fig. 4.35. Plot the downsampled and upsampled signal if the downsampling and upsampling factor is 2.

Solution The downsampling factor is 2. So discard samples at multiples of 2. The answer is shown below (Fig. 4.36).

While upsampling, we have to insert one zero between every sample. The output is shown below (Fig. 4.37).

Problem Prove that the following system is PR (Fig. 4.38).

Solution Let us assign names for the intermediate branches as shown in figure below (Fig. 4.39).

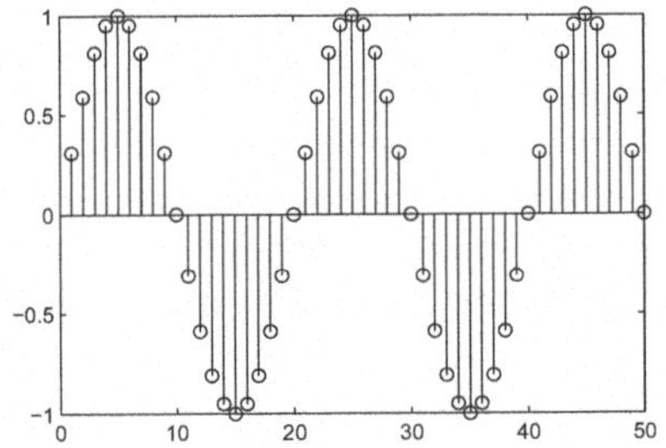

Fig. 4.35 Sinusoidal signal

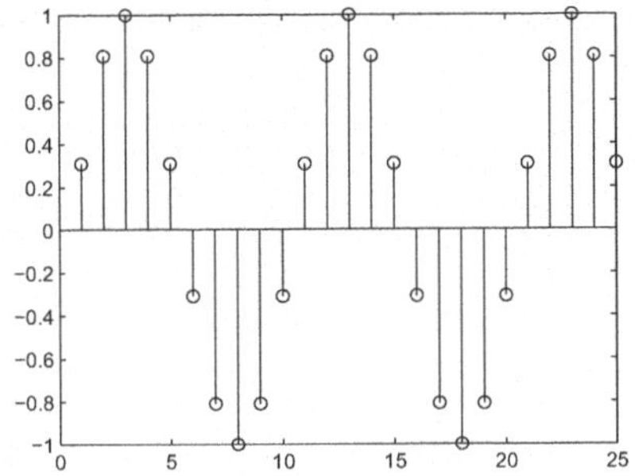

Fig. 4.36 Downsampled sinusoidal signal

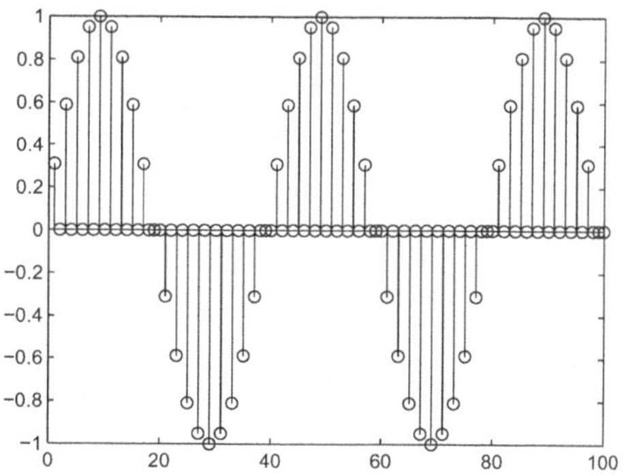

Fig. 4.37 Upsampled sinusoidal signal

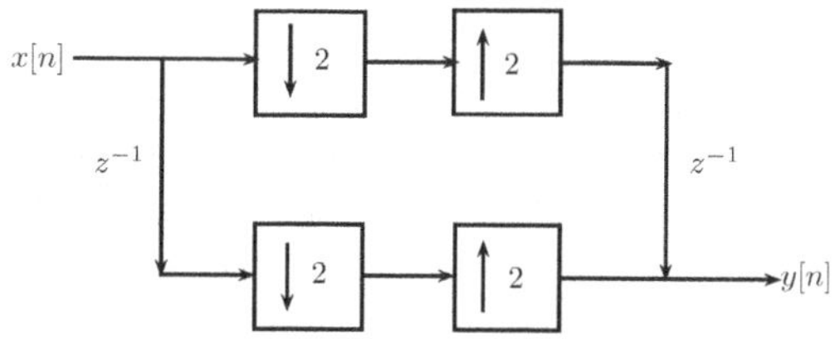

Fig. 4.38 Filter bank

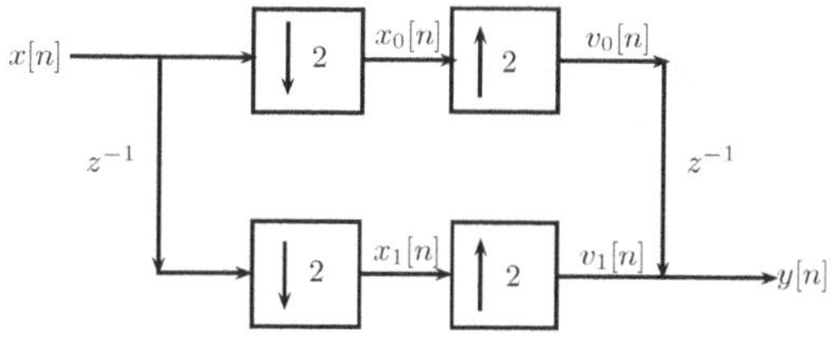

Fig. 4.39 Filter bank with intermediate signals

In order to prove that the system is PR, show that the output $y[n]$ is an amplified and delayed version of input $x[n]$. We can write the Z transform equations for each branch.

$$
\begin{aligned}
X_0[z] &= \frac{1}{2}\left\{X[z^{\frac{1}{2}}] + X[-z^{\frac{1}{2}}]\right\} \\
X_1[z] &= \frac{1}{2}\left\{z^{-\frac{1}{2}}X[z^{\frac{1}{2}}] + (-z)^{-\frac{1}{2}}X[-z^{\frac{1}{2}}]\right\} \\
V_0[z] &= X_0[z^2] \\
&= \frac{1}{2}\{X[z] + X[-z]\} \\
V_1[z] &= X_1[z^2] \\
&= \frac{1}{2}\left\{z^{-1}X[z] - z^{-1}X[-z]\right\} \\
Y[z] &= z^{-1}V_0[z] + V_1[z]
\end{aligned}
$$

$$= \frac{z^{-1}}{2}[2X[z]]$$
$$= z^{-1}X[z]$$
$$\Rightarrow y[n] = x[n-1]$$

The output $y[n]$ is the delayed version of $x[n]$. Hence, the system is PR.

Problem A piecewise constant function in the space V_1 is shown in Fig. 4.40. Find and plot the corresponding piecewise constant functions in V_0 and W_0 using Haar scaling and wavelet functions. Show the reconstruction process too.

Solution The piecewise constant function in V_0 is obtained by $\frac{1}{2}\{a[n] + a[n-1]\}$ if our function in V_1 is considered as $a[n]$. The output is shown in Fig. 4.41. Also note that the interval is of unit length.

The sequence on V_0 can be written as $\{8, 6, -5\}$. To plot the function on W_0, the equation is $\frac{1}{2}\{a[n] - a[n-1]\}$. The output is shown in Fig. 4.42.

Note that when the value is negative, the function will start from the negative cycle. The sequence on W_0 can be written as $\{-2, -2, 1\}$.

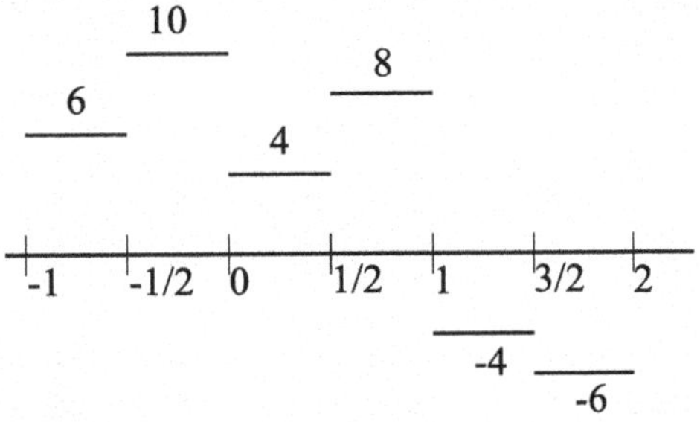

Fig. 4.40 Piecewise constant function in V_1

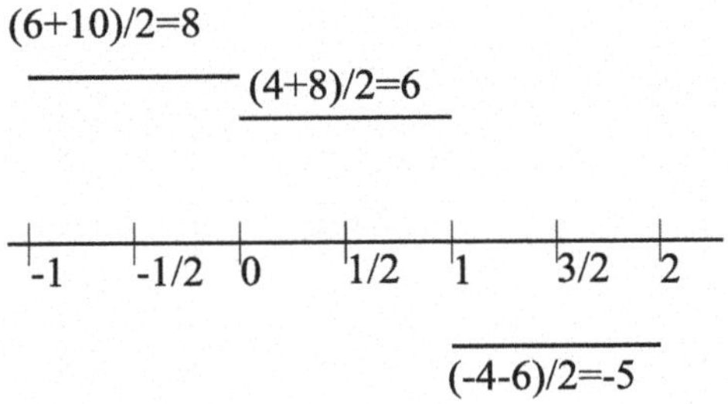

Fig. 4.41 Piecewise constant function in V_0

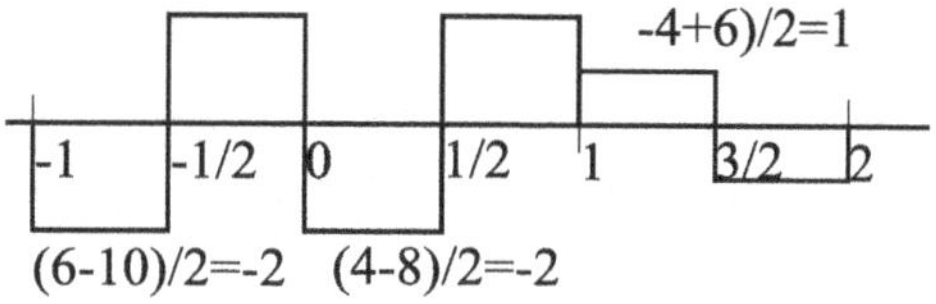

Fig. 4.42 Piecewise constant function in W_0

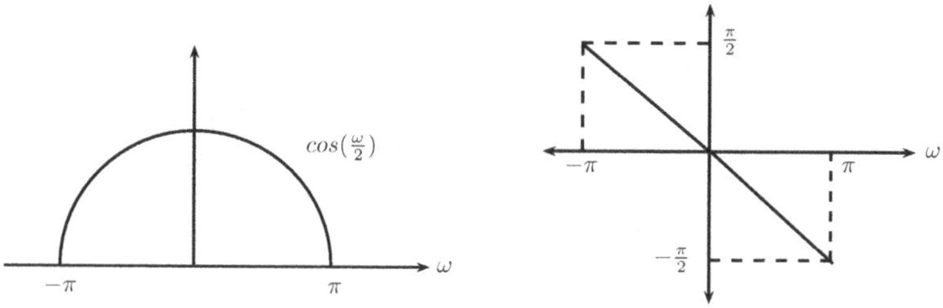

(a) Magnitude response (b) Phase response

Fig. 4.43 Low-pass filter response of Haar filter bank

For reconstruction, we have to first upsample the two sequences. So they become

$$\{8, 0, 6, 0, -5, 0\} \quad \text{and}$$
$$\{-2, 0, -2, 0, 1, 0\}$$

Now add and subtract the nonzero values and place alternatively.

$$\{8-2, 8+2, 6-2, 6+2, -5+1, -5-1\}$$
$$= \{6, 10, 4, 8, -4, -6\}$$

which is same as our original sequence.

Problem Prove that the filters $\frac{1}{2}[1 + z^-1]$ and $\frac{1}{2}[1 + z^-1]$ in the Haar filter bank structure are correspondingly low pass and high pass.

Solution Take the first filter $\frac{1}{2}[1 + z^{-1}]$. Its frequency response is $\frac{1}{2}[1 + e^{-j\omega}]$. Reducing further,

$$\frac{1}{2}[1 + e^{-j\omega}] = \frac{e^{-j\frac{\omega}{2}}}{2}[e^{j\frac{\omega}{2}} + e^{-j\frac{\omega}{2}}]$$
$$= e^{-j\frac{\omega}{2}} cos(\omega/2)$$

So the magnitude response is $| \; cos(\omega/2) \; |$, and the phase response is $(\omega/2)$. Plotting these two, we can see that this filter can be approximated as a low-pass filter with cutoff frequency $\frac{\pi}{2}$. But it is not an ideal low-pass filter. Also the phase is linear (Fig. 4.43).

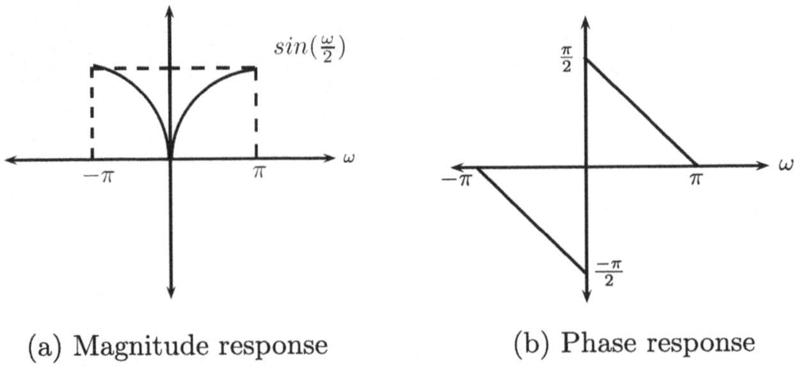

Fig. 4.44 High-pass filter response of Haar filter bank

Consider the next filter, $\frac{1}{2}[1 - z^{-1}]$. Its frequency response is $\frac{1}{2}[1 - e^{-j\omega}]$. Reducing further,

$$\begin{aligned}\frac{1}{2}[1 - e^{-j\omega}] &= \frac{e^{-j\frac{\omega}{2}}}{2}[e^{j\frac{\omega}{2}} - e^{-j\frac{\omega}{2}}]\\ &= \frac{e^{-j\frac{\omega}{2}}}{2} 2jsin(\omega/2)\\ &= e^{-j\frac{\omega}{2}} jsin(\omega/2)\end{aligned}$$

So the magnitude response is $\mid sin(\omega/2) \mid$, and the phase response is $(90 - \frac{\omega}{2})$. Plotting these two, we can see that this filter can be approximated as a high-pass filter with cutoff frequency $\frac{\pi}{2}$ (Fig. 4.44).

Chapter 5
Wavelet Design Using Regularity and Moments

In the last chapter, we discussed the relation of wavelet transform to filter banks. In this chapter, we will study in detail the scaling and wavelet functions of Haar wavelet and discuss in detail the properties of them. Then we will move upward to other classes of wavelets and will discuss in detail the steps to design a wavelet.

5.1 Haar Scaling Function

We had seen the Haar scaling and wavelet function in Fig. 4.24. The Haar scaling function $\phi(t)$ is nonzero only in the interval [0, 1].

$$\phi(t) = \begin{cases} 1; & 0 \leq t \leq 1 \\ 0; & elsewhere \end{cases} \tag{5.1}$$

The scaling function $\phi(t)$ is time limited and has finite energy. Now we will see the properties of this scaling function.

Orthogonality of Translates of $\phi(t)$

Assume that the scaling function is shifted to right by 1 unit. How will it look like? As shown in Fig. 5.1b, we know that this is $\phi(t-1)$ which is nonzero only in the interval [1, 2]. Now shift $\phi(t)$ to right by 2 units to get $\phi(t-2)$. $\phi(t-2)$ is nonzero in the interval [2, 3]. It is shown in Fig. 5.1c. Now if we shift $\phi(t)$ by k units to right, we get $\phi(t-k)$ which is nonzero in the interval k to $k+1$ as shown in Fig. 5.1d. Similarly, we can shift $\phi(t)$ to left and can obtain $\phi(t+1)$, $\phi(t+2)$ ... $\phi(t+k)$ as shown in Figs. 5.1e, f, and g.

M. S. Sinith et al., *Wavelets and Multi-Resolution Analysis*,
https://doi.org/10.1007/978-3-031-92750-8_5

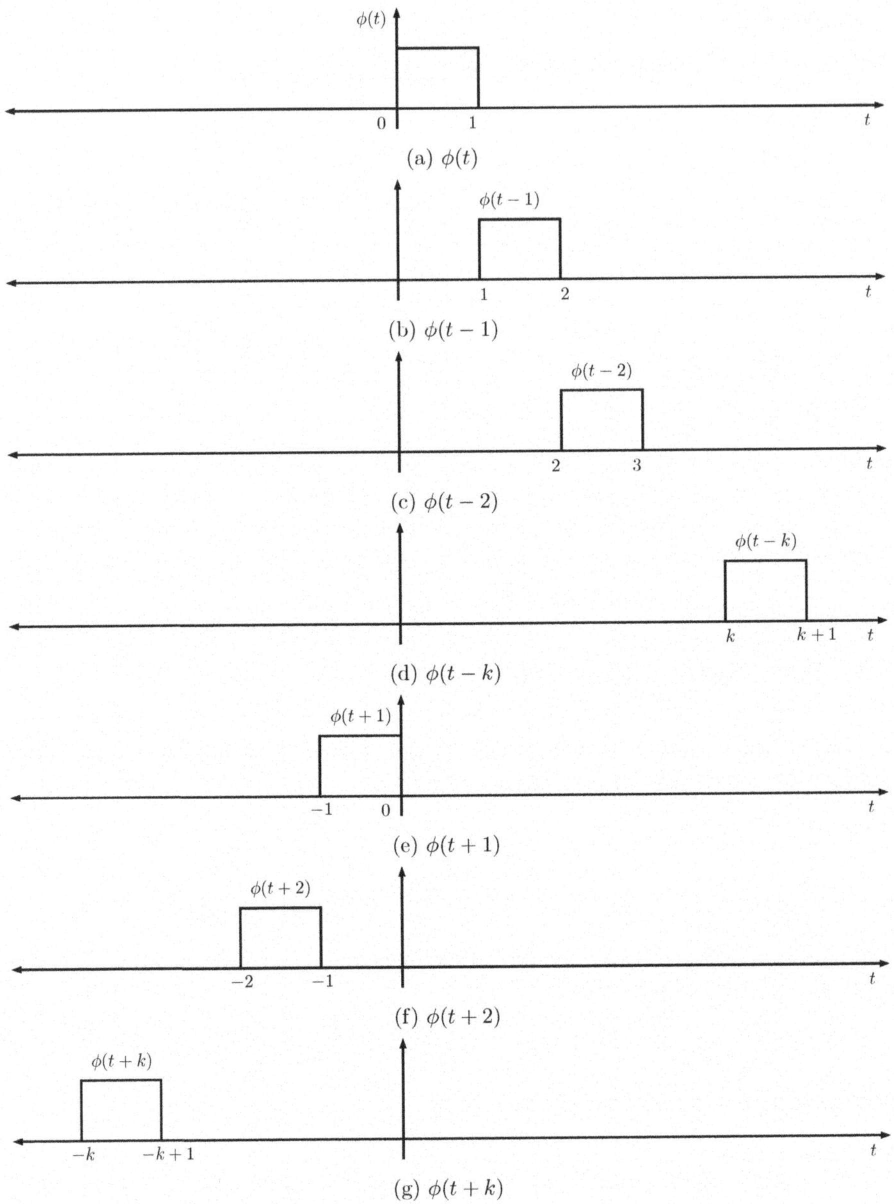

Fig. 5.1 Translates of scaling function $\phi(t)$

Now consider $\phi(5t - 3)$. How will it be? This is not easier as in our previous examples. We know $\phi(t)$ is nonzero in [0, 1], i.e., it is nonzero where the argument t in $\phi(t)$ falls in the range [0, 1]. In the same way, $\phi(5t - 3)$ will be nonzero where the argument $5t - 3$ falls in range [0, 1], i.e., when $5t - 3 = 0, t = 3/5$ and when

$5t-3=1, t=4/5$. So we can say $\phi(5t-3)$ is nonzero in the interval $[3/5, 4/5]$. In this way, we can find and draw any translated and scaled version of $\phi(t)$.

Now consider the integral $\int_{-\infty}^{\infty}\phi(t)\phi(t-1)$. From Fig. 5.1, we can see that $\phi(t)\phi(t-1)$ is zero as there is no overlap between them. The result is true for any shift $\phi(t)$ as there is no overlapping between $\phi(t)$ and its shifted versions. Also we can see that $\int_{-\infty}^{\infty}\phi(t)\phi(t) = 1$. The above results can be generalized to a single equation as

$$\int_{-\infty}^{\infty}\phi(t-m)\phi(t-n)dt = \delta_{m-n} \tag{5.2}$$

Equation 5.2 implies that the scaling function $\phi(t)$ is orthonormal to its own time shifts.

Vector Space Concept of $\phi(t)$

Moving on to vector spaces, now we have a set of orthonormal basis functions $\ldots, \phi(t+2), \phi(t+1), \phi(t), \phi(t-1), \phi(t-2), \ldots$. The space spanned by these basis functions is denoted by V_0.

$$V_0 = \underset{K}{Span}\{\overline{\phi(t-k)}\} \tag{5.3}$$

Any function $f(t)$ in the space V_0 can be written as $f(t) = \sum_{k=-\infty}^{\infty} a_k\phi(t-k)$, where a_ks are real numbers. We can generate the space V_0 by continuously varying a_ks. But the function $f(t)$ cannot be any arbitrary function of any shape. The space V_0 contains only those signals which are piecewise constant in the unit interval. (Refer section "Haar Filter Bank" for more details regarding piecewise linearity.)

Orthogonality of Translates of $\phi(2t)$

Now scale the function $\phi(t)$ by 2. We will get $\phi(2t)$. Shift these scaled functions as we had done before. How will they look like? Of course as shown in Fig. 5.2.

These functions are also nonoverlapping, and they are also orthogonal to each other (which is easy to prove in the same way we had done for $\phi(t)$). Now evaluate $\int \phi(2t)\phi(2t)$

$$\begin{aligned}\int \phi(2t)\phi(2t)dt &= \int_0^{\frac{1}{2}} 1.1dt \\ &= \frac{1}{2}\end{aligned} \tag{5.4}$$

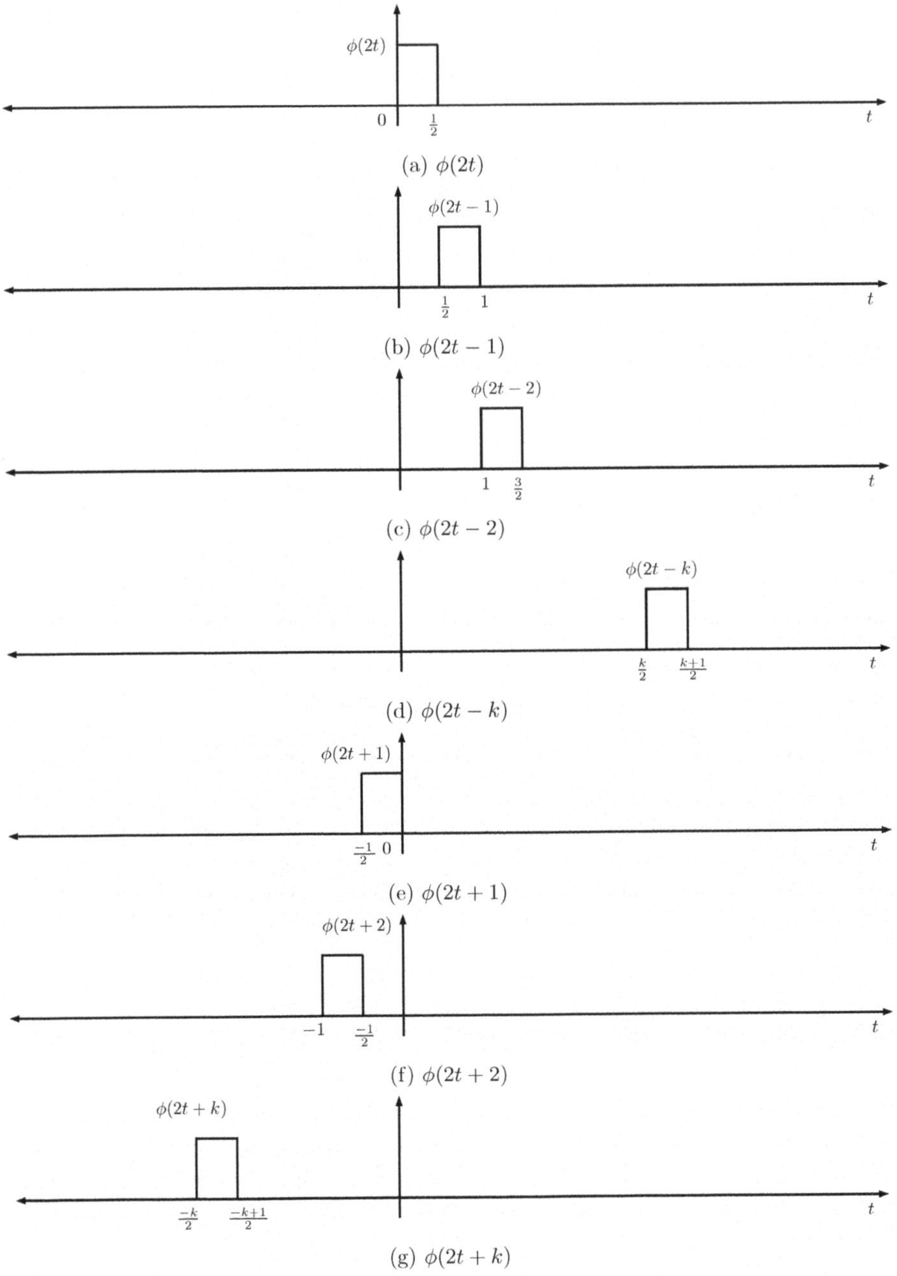

Fig. 5.2 Translates of scaled scaling function $\phi(2t)$

The bases function in V_1 is orthogonal but not orthonormal as the above integral is not equal to 1. How can we make the bases orthonormal? Multiply each $\phi(t)$ by $\sqrt{2}$. Now the integral becomes

$$\begin{aligned} \int \sqrt{2}\phi(2t)\sqrt{2}\phi(2t)dt &= \int_0^{\frac{1}{2}} \sqrt{2}.\sqrt{2}dt \\ &= 1 \end{aligned} \tag{5.5}$$

Vector Space Concept of $\phi(2t)$

The vector space concept here is an extended version of the previous one. Here the basis functions are $\ldots, \phi(2t+2), \phi(2t+1), \phi(2t), \phi(2t-1), \phi(2t-2), \ldots$. They span the space V_1. Any function $f_1(t)$ in the space V_1 can be written as $f(t) = \sum_{k=-\infty}^{\infty} a_k \phi(2t-k)$, where a_ks are real numbers. These functions should be piecewise constant in half intervals.

$$V_1 = \underset{K}{Span}\{\overline{\sqrt{2}\phi(2t-k)}\} \tag{5.6}$$

Now consider V_2. The space V_2 is spanned by the translates of $\phi(4t)$ which will be nonzero in the interval $[0, 1/4]$. This function $\phi(4t)$ is also nonoverlapping with its translates and hence is orthogonal. Now we can check for normalization.

$$\begin{aligned} \int \phi(4t)\phi(4t)dt &= \int_0^{\frac{1}{4}} 1.1dt \\ &= \frac{1}{4} \end{aligned} \tag{5.7}$$

Hence, it is not the normalized one. So to normalize it, multiply each $\phi(t)$ by $\sqrt{4} = 2$. So

$$\begin{aligned} \int 2\phi(2t)2\phi(2t)dt &= \int_0^{\frac{1}{4}} 2.2dt \\ &= 1 \end{aligned} \tag{5.8}$$

The space V_2 is now spanned by

$$V_2 = \underset{K}{Span}\{\overline{2\phi(4t-k)}\} \tag{5.9}$$

In general, for a jth space V_j,

$$
\begin{aligned}
V_j &= \underset{K}{Span}\overline{\{\sqrt{2^j}\phi(2^j t - k)\}} \\
V_j &= \underset{K}{Span}\overline{\{2^{j/2}\phi(2^j t - k)\}}
\end{aligned}
\tag{5.10}
$$

As we had seen in section "Haar Filter Bank", V_1 is a finer space than V_0 or it contains more details. It is interesting to note that any function in V_0 can be expressed in terms of the bases of V_1 and the converse is not true. How is it possible? Because

$$
\begin{aligned}
\phi(t) &= \phi(2t) + \phi(2t - 1) \\
\phi(2t) &= \phi(4t) + \phi(4t - 1)
\end{aligned}
\tag{5.11}
$$

and so on. So we can say that V_0 is contained in V_1. What is missing in V_0 compared to V_1? It is the wavelet function in the space W_0, and this will be discussed in the next section.

5.2 Haar Wavelet Function

Haar wavelet function $\psi(t)$ is shown in Fig. 4.24b. It is given by

$$
\psi(t) = \begin{cases} 1; & 0 \le t \le \frac{1}{2} \\ -1; & \frac{1}{2} \le t \le 1 \\ 0; & elsewhere \end{cases}
\tag{5.12}
$$

Like the scaling function, we can easily prove that the wavelet function is orthonormal as it does not overlap with its translates. The translates of $\psi(t)$ are shown in Fig. 5.3.

So we can write

$$
\int_{-\infty}^{\infty} \psi(t - m)\psi(t - n)dt = \delta_{m-n}
\tag{5.13}
$$

The vector space spanned by the wavelet function is defined as W_0 and is given by

$$
W_0 = \underset{K}{Span}\overline{\{\psi(t - k)\}}
\tag{5.14}
$$

How will a function in W_0 look like? There are two options. One is that it should be constant in the first half of unit interval with a value $+x$ units and should also be

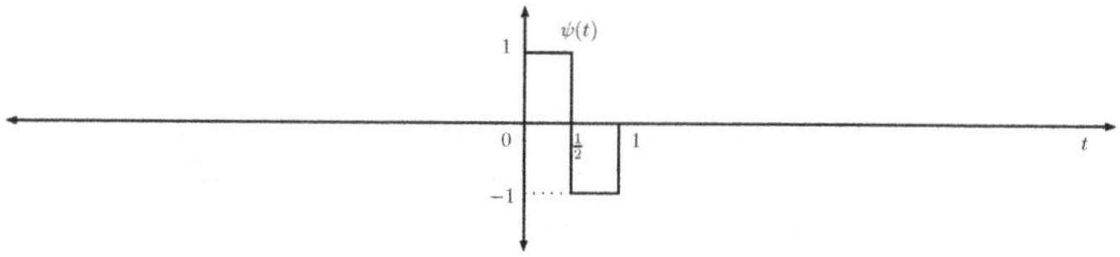

(a) $\psi(t)$

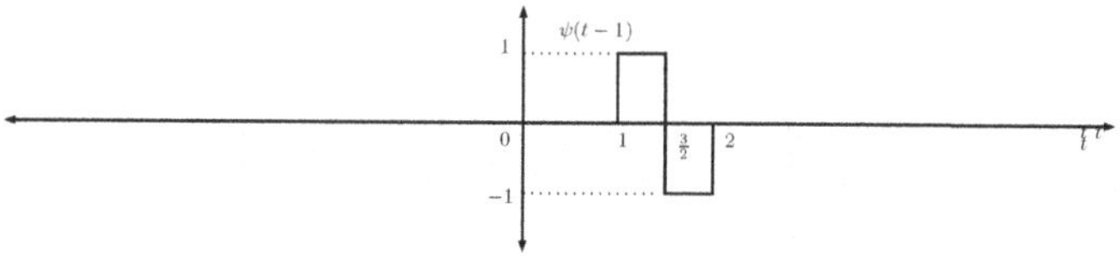

(b) $\psi(t-1)$

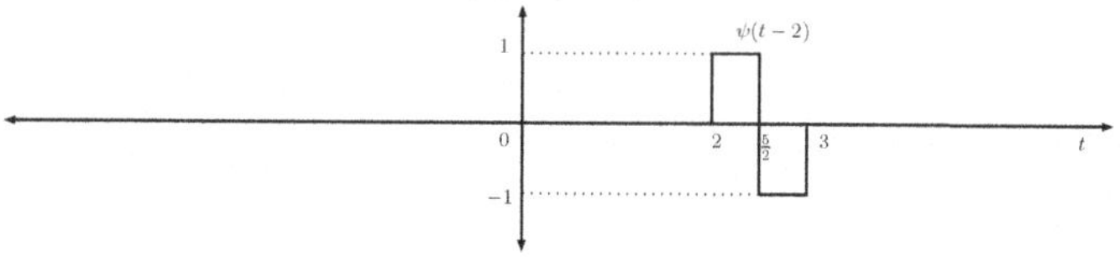

(c) $\psi(t-2)$

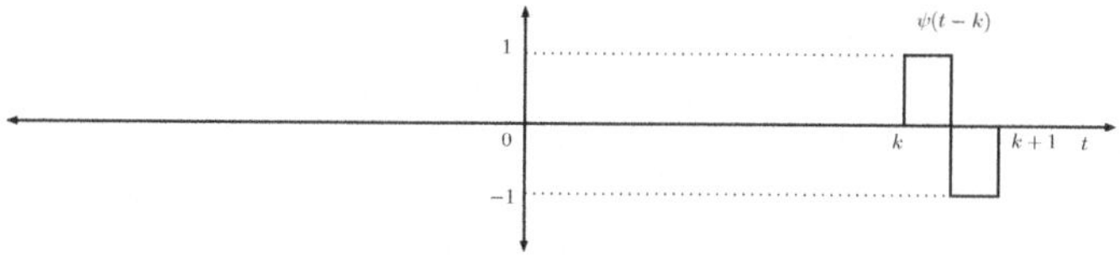

(d) $\psi(t-k)$

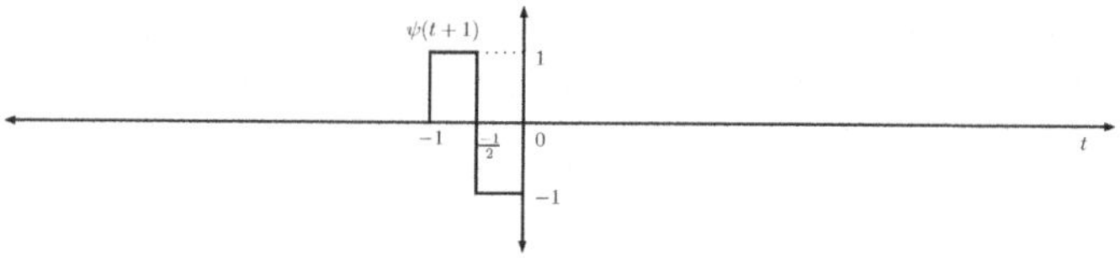

(e) $\psi(t+1)$

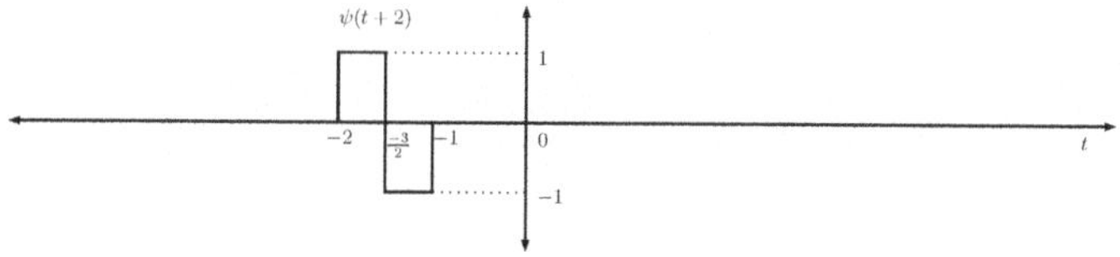

(f) $\psi(t+2)$

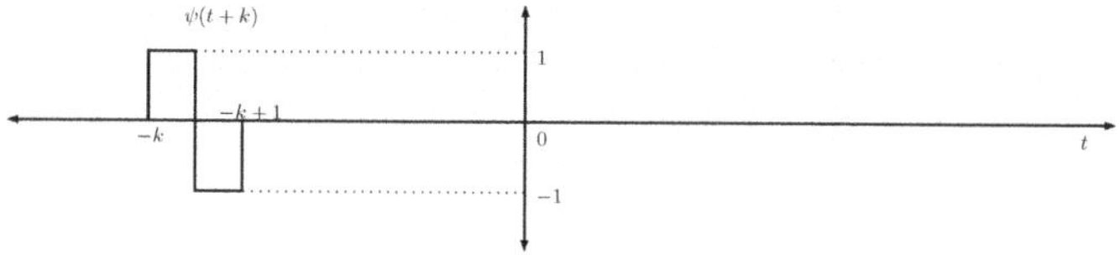

(g) $\psi(t+k)$

Fig. 5.3 Translates of wavelet function $\psi(t)$

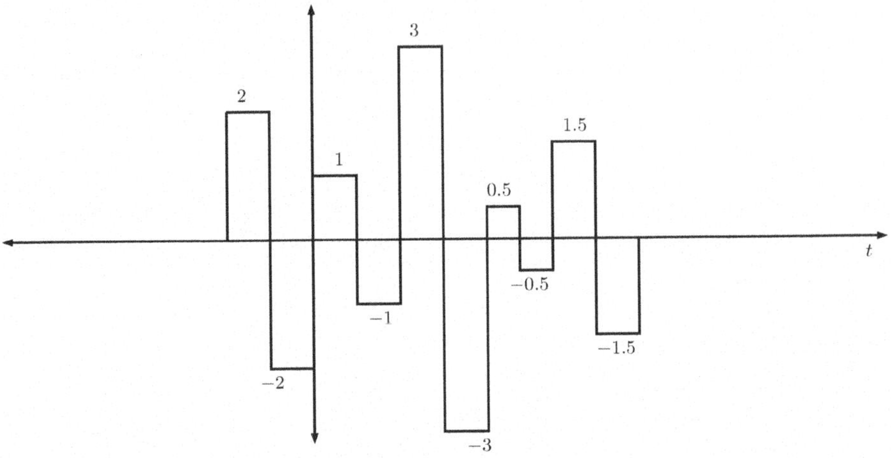

Fig. 5.4 Example of a function $f_1(t)$ in W_0

constant in the next half interval with $-x$. Or, it should be constant in the first half of unit interval with a value $-x$ units and should also be constant in the next half interval with $+x$. Such a function $f_1(t)$ is shown in Fig. 5.4.

Now move on to the scaled version of $\psi(t)$. In the scaling function space, V_0 is contained in V_1. Let us check whether it is the same for wavelet function. The scaled wavelet function and its translates are shown in Fig. 5.5. These functions are also nonoverlapping, and they are also orthogonal to each other. Hence, we can write

$$\int_{-\infty}^{\infty} \psi(2t-m)\psi(2t-n)dt = \delta_{m-n} \tag{5.15}$$

Here the basis functions are $\ldots, \psi(2t+2), \psi(2t+1), \psi(2t), \psi(2t-1), \psi(2t-2), \ldots$. They span the space W_1.

$$W_1 = \underset{K}{Span}\{\overline{\psi(2t-k)}\} \tag{5.16}$$

Can we represent the function $f_1(t)$ (shown in Fig. 5.4) in terms of $\psi(2t)$ and its translates? The answer is NO. Hence, no function in W_0 can be expressed in terms of the bases of W_1. So W_0 is not contained in W_1. Now evaluate the

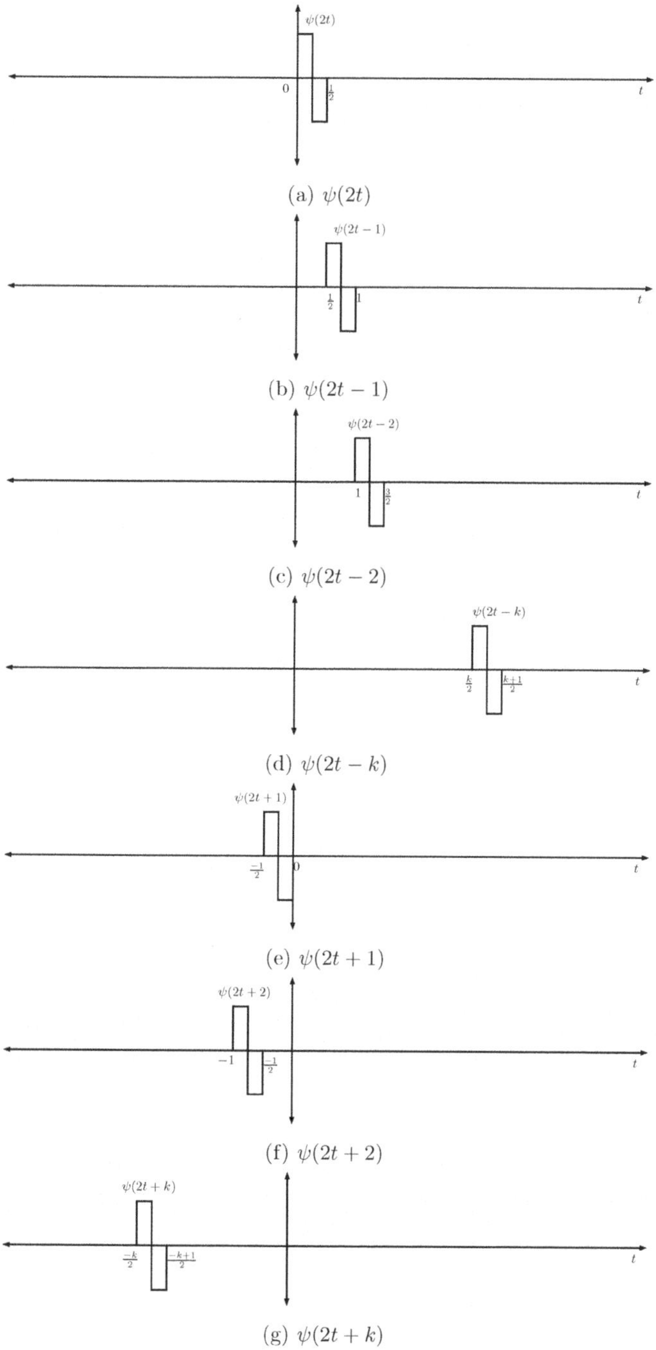

Fig. 5.5 Translates of scaled wavelet function $\psi(2t)$

integral $\int \psi(t)\psi(2t)dt$. What is the result? Of course, it is 0. Now what will be $\int \psi(t)\psi(2t-1)dt$? Again the answer is 0.

$$
\begin{aligned}
\int \psi(t)\psi(2t)dt &= 0 \\
\int \psi(t)\psi(2t-1)dt &= 0 \\
&\vdots \\
\int \psi(t)\psi(2t-k)dt &= 0
\end{aligned}
\tag{5.17}
$$

This implies that the space W_0 is orthogonal to W_1. In the same way, we can prove that W_1 is orthogonal to W_2. Thus we had derived two important results in wavelet theory. They are:

- The space spanned by the bases functions of scaling function is nested, i.e.,

$$V_{-\infty} \subset \ldots \subset V_{-1} \subset V_0 \subset V_1 \subset \ldots V_{\infty}.$$

- The space spanned by the bases functions of wavelet function is orthogonal, i.e.,

$$W_{-\infty} \perp \ldots \perp W_{-1} \perp W_0 \perp W_1 \perp \ldots W_{\infty}.$$

Earlier we had pointed that there is something missing in V_0 making it a subset of V_1. In order to find this, let us consider a function $f_2(t)$ in the space V_1 as shown in Fig. 5.6.

$$f_2(t) = 5\phi(2t) + 1\phi(2t-1) \tag{5.18}$$

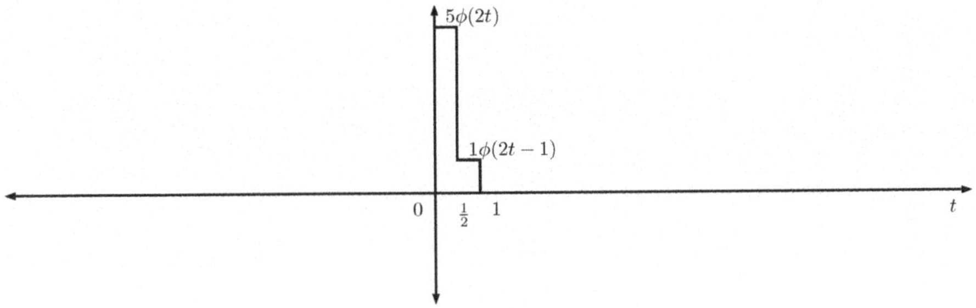

Fig. 5.6 Example of a function $f_2(t)$ in V_1

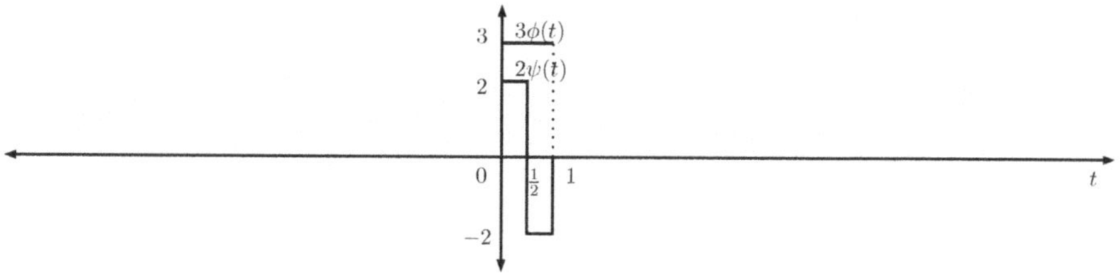

Fig. 5.7 Example of a function $f_2(t)$ in V_1

This function can be rewritten as

$$\begin{aligned} f_2(t) &= 5\phi(2t) + 1\phi(2t-1) \\ &= \frac{5+1}{2}\phi(t) + \frac{5-1}{2}\psi(t) \\ &= 3\phi(t) + 2\psi(t) \end{aligned} \tag{5.19}$$

The above concept is shown in Fig. 5.7.

So a signal in V_1 can be represented in terms of V_0 and W_0. Mathematically, it can be written as

$$V_1 = V_0 \oplus W_0 \tag{5.20}$$

Now it is clear that W_0 is missing in V_0 making it a subset of V_1. The result can be extended as

$$V_2 = V_1 \oplus W_1 \tag{5.21}$$

In general,

$$V_j = V_{j-1} \oplus W_{j-1} \tag{5.22}$$

But, $V_{j-1} = V_{j-2} \oplus W_{j-2}$. Therefore,

$$\begin{aligned} V_j &= V_{j-2} \oplus W_{j-2} \oplus W_{j-1} \\ &\vdots \quad \vdots \quad \vdots \\ V_j &= V_0 \oplus W_0 \oplus \ldots \oplus W_{j-3} \oplus W_{j-2} \oplus W_{j-1} \end{aligned} \tag{5.23}$$

The above equation shows that any signal in V_j can be expressed in terms of the bases in V_0, W_0 ... W_{j-1}. The process is shown schematically in Fig. 5.8.

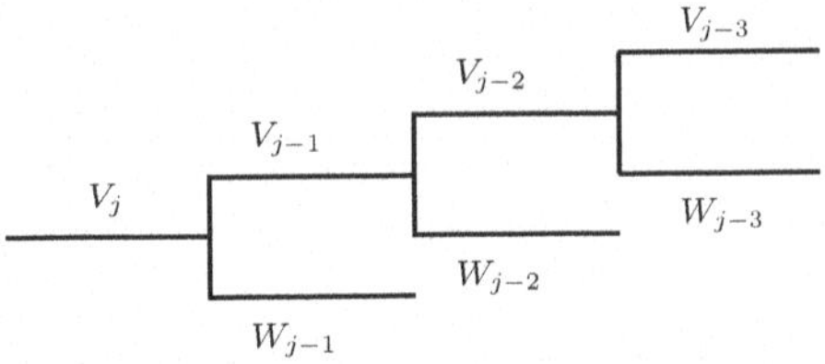

Fig. 5.8 Figure showing wavelet decomposition

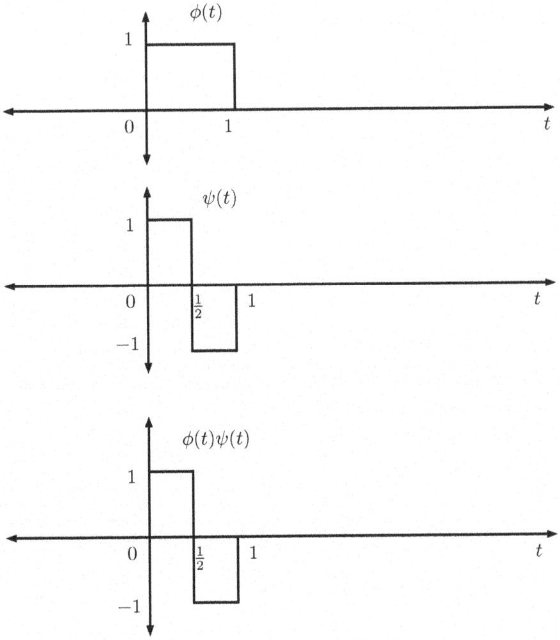

Fig. 5.9 Orthogonality of $\phi(t)$ and $\psi(t)$

5.3 Relation Between Scaling and Wavelet Functions

Consider the scaling function in V_0 and the wavelet function in W_0, i.e., $\phi(t)$ and $\psi(t)$. They do overlap in the interval [0, 1]. Now evaluate the integral $\int_t \phi(t)\psi(t)dt$.

$$\int_t \phi(t)\psi(t)dt = \int_0^{0.5} 1.1dt + \int_{0.5}^{1} 1.(-1)dt = 0 \tag{5.24}$$

Hence, it is clear that $\phi(t)$ is orthogonal to $\psi(t)$. The concept is shown in Fig. 5.9.

5.4 Refinement Relations

Equation 5.11 is with respect to unnormalized bases. Now we go for normalized bases. So

$$\phi(t) = \frac{1}{\sqrt{2}}\sqrt{2}\phi(2t) + \frac{1}{\sqrt{2}}\sqrt{2}\phi(2t-1) \tag{5.25}$$

It can also be derived that

$$\phi(t) = \phi(2t) - \phi(2t-1) \tag{5.26}$$

which can also be normalized to

$$\psi(t) = \frac{1}{\sqrt{2}}\sqrt{2}\phi(2t) - \frac{1}{\sqrt{2}}\sqrt{2}\phi(2t-1) \tag{5.27}$$

The above equations can be rearranged as

$$\begin{aligned}\phi(t) &= \sum_{k=0}^{1} h(k)\sqrt{2}\phi(2t-k)\\ \psi(t) &= \sum_{k=0}^{1} g(k)\sqrt{2}\phi(2t-k)\end{aligned} \tag{5.28}$$

where $h(k)$ is the low-pass filter coefficient and $g(k)$ is the high-pass filter coefficient in the filter bank structure.

In general, the refinement relations are

$$\begin{aligned}\phi(t) &= \sum_{k} h(k)\sqrt{2}\phi(2t-k)\\ \psi(t) &= \sum_{k} g(k)\sqrt{2}\phi(2t-k)\end{aligned} \tag{5.29}$$

5.5 Daubechies Wavelets

Till now we discussed in detail about Haar scaling and wavelet function, and we designed the Haar filter bank in the last chapter. Haar multiresolution analysis has one low-pass filter and one high-pass filter in both analysis and synthesis sides. Now we are going to higher families of wavelets. This indirectly indicates that Haar wavelet is inadequate for many applications or it has some drawbacks. What are they? Assume that we have a constant sequence given by $x[n] = c_1 \quad \forall \quad n$. What

Fig. 5.10 Haar analysis filter bank with x[n] as input

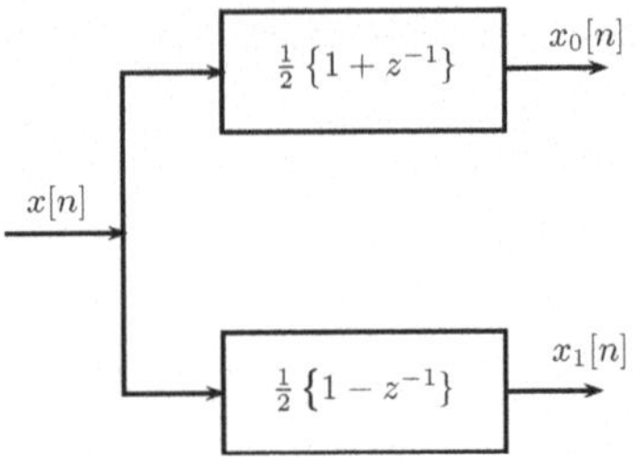

will happen when this is given as input for a Haar filter bank? We know the low-pass filter transfer function is given by $H_0[z] = \frac{1+z^{-1}}{2}$ and high-pass filter transfer function is given by $H_1[z] = \frac{1-z^{-1}}{2}$ (Fig. 5.10). When the signal $x[n] = c_1$ is applied, what will be the outputs from two branches? In Z domain, we can write it as simple multiplication between input and transfer function, i.e., $X_0[z] = X[z]H_0[z]$ and $X_1[z] = X[z]H_1[z]$.

Consider the high-pass filter output $X_1[z]$.

$$X_1[z] = X[z]H_1[z] = X[z]\frac{1}{2}\{1 - z^{-1}\}$$

In time domain, $x_1[n] = \frac{x[n]-x[n-1]}{2}$. This is equal to 0, as our input signal $x[n]$ is a constant for all n. So in the case of Haar, the constant part of the input signal is destroyed in the high-pass section. That is why we are looking forward to other wavelet categories. As we will see in the subsequent sections, some other wavelets have better performance.

Daubechies Filter Bank

The next higher family of Haar is Daubechies family, which was created by Ingrid Daubechies. The first member of this family is Daubechies 2 wavelet. Daubechies 2 wavelet has length 4 filters in its filter bank structure. As we move up in the family, the length of filter will increase by 2, i.e., Daubechies 3 has length 6 filters, Daubechies 4 has length 8 filters, and so on. Haar can be considered as Daubechies 1 wavelet.

Now we will discuss how to design Daubechies 2 filter coefficients from Haar filter bank. Haar filter bank has one $(1 - z^{-1})$ term in its high-pass filter. In order to design the next higher member, i.e., the Daubechies 2 wavelet, we should cascade one more $(1 - z^{-1})$ term. Every $(1 - z^{-1})$ term reduces a polynomial at the input to one degree lower. For example, in Daubechies 2, we have two $(1 - z^{-1})$ terms. Assume that the input is a polynomial of degree one, i.e., $x[n] = 3n + 5$. The output of first $(1 - z^{-1})$ term will be $x[n] - x[n - 1] = 3n + 5 - (3(n - 1) + 5) = 3$. This constant term will be destroyed in the next $(1 - z^{-1})$ term as discussed in Haar case. As we move to higher members, we have more and more $(1 - z^{-1})$ terms in the high-pass section.

Daubechies 2 family has length 4 filters. We have two $(1-z^{-1})$ terms in the high-pass section and correspondingly two $(1+z^{-1})$ terms in the low-pass filter. We need a low-pass filter whose impulse response is like $H_0[z] = h_0+h_1z^{-1}+h_2z^{-2}+h_3z^{-3}$ as the length of the filter is 4. There are three zeros. Already we have two zeros from $[1 + z^{-1}]^2$ term. We need one more zero and let it be B_0. So

$$\begin{aligned} H_0[z] &= C_0(1 + z^{-1})^2(1 + B_0z^{-1}) \\ &= C_0(1 + 2z^{-1} + z^{-2})(1 + B_0z^{-1}) \\ &= C_0(1 + 2z^{-1} + z^{-2} + B_0z^{-1} + 2B_0z^{-2} + B_0z^{-3}) \\ &= C_0(1 + (2 + B_0)z^{-1} + (1 + 2B_0)z^{-2} + B_0z^{-3}) \end{aligned} \tag{5.30}$$

where C_0 is a constant. Comparing with the standard form,

$$\begin{aligned} h_0 &= 1 \\ h_1 &= 2 + B_0 \\ h_2 &= 1 + 2B_0 \\ h_3 &= B_0 \end{aligned} \tag{5.31}$$

So what we have to do is to find out the value of B_0. We know the impulse response is orthogonal to its even shifts (Fig. 5.11). So we can write

$$h_0h_2 + h_1h_3 = 0 \tag{5.32}$$

This is a nontrivial equation.
Substituting values for coefficients from Eq. (5.31),

$$\begin{aligned} &h_0h_2 + h_1h_3 = 0 \\ &1.(1 + 2B_0) + (2 + B_0)B_0 = 0 \\ &1 + 2B_0 + 2B_0 + B_0^2 = 0 \\ &1 + 4B_0 + B_0^2 = 0 \end{aligned} \tag{5.33}$$

h_0	h_1	h_2	h_3		
		h_0	h_1	h_2	h_3

Fig. 5.11 Figure showing even shifts of $H_0[z]$

Solving the quadratic equation in the normal way, we get $B_0 = -2+\sqrt{3}$ and $-2-\sqrt{3}$. From these two values which value should we select? Observing them, we can see that one is inside the unit circle and the other outside. $-2-\sqrt{3}$ is outside the unit circle and $-2+\sqrt{3}$ inside. Choose the root inside the unit circle for minimum phase condition. So $B_0 = -2+\sqrt{3}$. Impulse response coefficients are now $[1, 2+B_0, 1+2B_0, B_0] = [1, \sqrt{3}, -3+2\sqrt{3}, -2+\sqrt{3}]$. Now we need to find the value of constant C_0.

From Eq. 4.18, we know for a PR system, $T[z] = cz^{-n_0}$

$$\begin{aligned} T[z] &= F_0[z]H_0[z] + F_1[z]H_1[z] \\ &= H_1[-z]H_0[z] + H_0[-z]H_1[z] \\ &= cz^{-n_0} \end{aligned} \tag{5.34}$$

In Haar case, $H_1[z] = H_0[-z]$. In general, we can say $H_1[z]$ should be related to $H_0[-z]$. Let us choose $H_1[z] = z^{-n_0}H_0[-z^{-1}]$. The selection is not arbitrary. There are reasons for this selection. z^{-n_0} term is to introduce causality. How this term ensures causality?

In Haar case, $H_0[z] = \frac{1+z^{-1}}{2}$, $H_0[z^{-1}] = \frac{1+z}{2}$. So $z^{-n_0}H_0[-z^{-1}] = z^{-n_0}\left[\frac{1-z}{2}\right]$. The presence of term z makes the filter noncausal. So for making it causal, $n_0 = 1$. Thus we get $H_1[z] = z^{-1}\left[\frac{1-z}{2}\right] = \frac{z^{-1}-1}{2}$. n_0 is always chosen as an odd number because the length of the filter is an even number, and to make it causal, n_0 should be odd. Thus z^{-n_0} term makes the filter causal. The z^{-1} factor is to bring power complimentary property, and "-" sign is to change the low-pass filter $H_0[z]$ to high-pass filter. These justify why we choose $H_1[z] = z^{-n_0}H_0[-z^{-1}]$. Substituting this condition in the above equation

$$\begin{aligned} T[z] &= H_1[-z]H_0[z] + H_0[-z]H_1[z] \\ cz^{-n_0} &= (-1)^{n_0}z^{-n_0}H_0[z^{-1}]H_0[z] - H_0[-z]z^{-n_0}H_0[-z^{-1}] \\ c &= (-1)^{n_0}H_0[z]H_0[z^{-1}] - H_0[-z]H_0[-z^{-1}] \\ c &= -H_0[z]H_0[z^{-1}] - H_0[-z]H_0[-z^{-1}] \quad \text{as } n_0 \text{ is odd} \end{aligned} \tag{5.35}$$

If $K_0[z] = H_0[z]H_0[z^{-1}]$, $K_0[-z] = H_0[-z]H_0[-z^{-1}]$. So the above equation gets modified as

$$K_0[z] + K_0[-z] = \text{constant} \tag{5.36}$$

Thus in general, the sequence corresponding to $K_0[z]$ is the square of the norm of the impulse response which should be equal to 1. Apply the same in our Daubechies wavelet design. Hence, choose C_0 in our design process so that $C_0^2\{h_0^2 + h_1^2 + h_2^2 +$

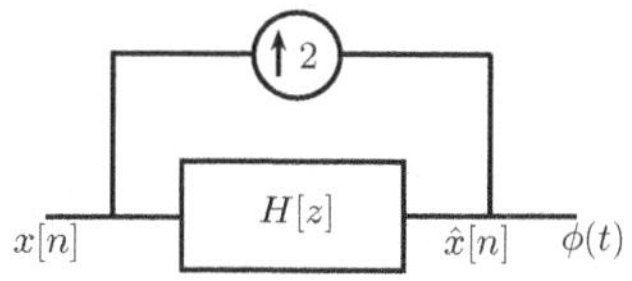

Fig. 5.12 Iterative method to generate scaling function

$h_3^2\} = 1$. We obtained the values of h_0 *to* h_3, and substituting these values, we get

$$C_0^2[1^2 + \sqrt{3}^2 + (-3 + 2\sqrt{3})^2 + (-2 + \sqrt{3})^2] = 1$$

Solving, we get $C_0 = 0.4829$. Multiplying with this constant, the values of low-pass filter coefficients are obtained as

$$[h_0 \quad h_1 \quad h_2 \quad h_3] = [0.483 \quad 0.8365 \quad 0.2241 \quad -0.1294]$$

Remember that $h_k s$ are normalized filter coefficients and the scaling function coefficient $c_k = \sqrt{2}h_k$. So $c_k s$ for Daubechies 2 filter is

$$[0.683 \quad 1.183 \quad 0.317 \quad -0.183]$$

Daubechies 2 Scaling and Wavelet Functions

There are different methods to plot the scaling function once we know the scaling function coefficients. A simple method is illustrated in the following figure (Fig. 5.12).

Initially, $x[n]$ is an impulse signal. This is given as the input to the filter $H[z]$ whose coefficients are known to us. The output is upsampled by a factor of 2 and is given as the input to the filter again. This process is repeated 5 or 6 times, and the scaling function is obtained. A simple MATLAB program for the process is shown below and the output is shown in Fig. 5.13a.

MATLAB Code

```
%.....................................................................
% ...........Program to plot Daubechies 2 scaling function..............
h=[0.683 1.183 0.317 -0.183]; %dB2 filter coefficients
ip1=1;%impulse
for i=1:6
    opin=conv(h,ip1); %filter input is the upsampled output
    ip1=upsample(opin,2);%upsampling by 2
end;
plot(opin,'k','LineWidth',2);
title('dB2 Scaling Function');
```

```
xlabel('Time');
ylabel('Amplitude');
%Output is shown in figure 5.13(a)
```

Now we have the low-pass filter coefficients. The high-pass filter coefficients can be obtained by just reversing them and changing the sign of alternate values. The MATLAB code is given below. The output is shown in Fig. 5.13b.

MATLAB Code

```
%...........................................................
%...........Program to plot Daubechies 2 wavelet function...............
h=[0.683 1.183 0.317 -0.183];
k=4;        %no: of coefficients
for m=1:4
   g(k)=((-1)^(1-k))*h(m);%High pass filter coefficients
   k=k-1;
end
ip2=1;
op1=conv(g,ip2);
op2=upsample(op1,2);
for i=1:7
    op2in=conv(h,op2); %filter input is the upsampled output
    op2=upsample(op2in,2);%upsampling by 2
end;
plot(op2in,'k','LineWidth',2);
title('dB2 Wavelet Function');
xlabel('Time');
ylabel('Amplitude');
%Output is shown in figure 5.13(b)
```

The convolution process may not converge always. This can be explained on the basis of regularity. The filters need to obey a property called regularity for the iterative convolution process to converge to a function which is continuous or almost continuous in a finite number of points. One way is to introduce more and more $(1 + z^{-1})$ terms or in other words to introduce more zeros in the point $z = -1$. Correspondingly, more and more $(1 - z^{-1})$ terms will be introduced in the high-pass filter side, i.e., more and more zeros in $z = 1$. A filter is said to be K regular if its Z transform has K zeros at $z = e^{j\pi}$. As the seniority in the Daubechies family increases, regularity also increases.

Properties of Scaling Function and Wavelet Function Coefficients

The above-described process becomes difficult as the length of the filter increases. It will be easier if we have a clear procedure for the design. So here we will discuss some properties that the scaling and wavelet function coefficients should satisfy based on which we can design them. We will see them one by one. We had seen these

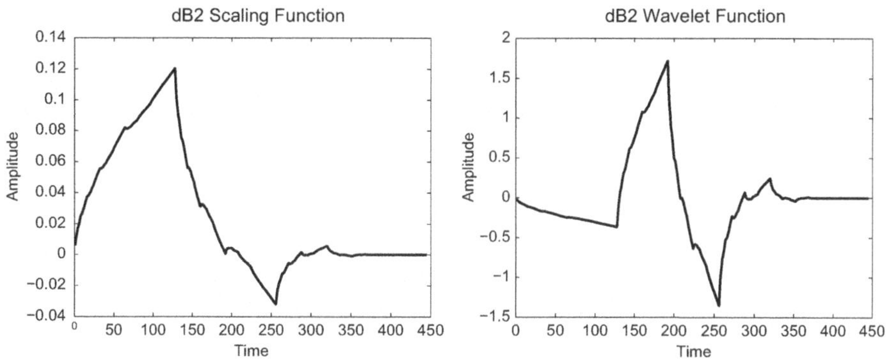

(a) Daubechies 2 scaling function $\phi(t)$ (b) Daubechies 2 wavelet function $\psi(t)$

Fig. 5.13 Daubechies 2 scaling and wavelet functions

properties in the scaling and wavelet functions of Haar. Now we are generalizing the idea to get a design procedure. Before that we will once again have a look at the refinement relations. They are

$$\begin{aligned} \phi(t) &= \sum_{k=0}^{N-1} c_k \phi(2t-k) = \sum_{k=0}^{N-1} h_k \sqrt{2} \phi(2t-k) \quad \text{as } c_k = h_k\sqrt{2} \\ \psi(t) &= \sum_{k=0}^{N-1} d_k \phi(2t-k) = \sum_{k=0}^{N-1} g_k \sqrt{2} \phi(2t-k) \quad \text{as } d_k = g_k\sqrt{2} \end{aligned} \tag{5.37}$$

We are mainly concentrating on scaling function coefficients on the design. Why? Look at the refinement relations. Once we know $\phi(2t)$, we can get $\phi(t)$ and $\psi(t)$. Also the wavelet function coefficients g_ks can be obtained from h_ks as described in the following section. So the knowledge of scaling function coefficients is sufficient for the design of whole wavelet. We will see in the following section the properties to be satisfied by the scaling function coefficients.

Property 1: Normalization of Scaling Function Coefficients
This property depends on the type of normalization used for scaling function $\phi(t)$. Usually, the normalization used is $\int_{-\infty}^{\infty} \phi(t)dt = 1$. In other words, we can say, there should be unit area under the scaling function.

$$\begin{aligned} \int_{-\infty}^{\infty} \phi(t)dt &= \int_{-\infty}^{\infty} \sum_{k=0}^{N-1} h_k \sqrt{2} \phi(2t-k)dt \\ &= \sum_{k=0}^{N-1} h_k \sqrt{2} \int_{-\infty}^{\infty} \phi(2t-k)dt \end{aligned} \tag{5.38}$$

Let $2t - k = x$. Therefore $dt = \frac{1}{2}dx$.

$$\int_{-\infty}^{\infty} \phi(t)dt = \sum_{k=0}^{N-1} h_k \sqrt{2}\frac{1}{2} \int_{-\infty}^{\infty} \phi(x)dx \tag{5.39}$$

$\int_{-\infty}^{\infty} \phi(t)dt = \int_{-\infty}^{\infty} \phi(x)dx$ as they are only the matter of variables.
So

$$\begin{aligned} 1 &= \sum_{k=0}^{N-1} h_k \sqrt{2}\frac{1}{2} \\ \Rightarrow & \boxed{\sum_{k=0}^{N-1} h_k = \sqrt{2}} \end{aligned} \tag{5.40}$$

Property 2: Orthonormality of Scaling Function to Its Own Translates
We had seen in Haar that the scaling function is orthonormal to its own translates. It is given in Eq. 5.2.

$$\int_{-\infty}^{\infty} \phi(t - m)\phi(t - n)dt = \delta_{m-n}$$

Using refinement relations, this equation can be modified as

$$\begin{aligned} \int_{-\infty}^{\infty} \phi(t - m)\phi(t - n)dt &= \int_{-\infty}^{\infty} \sum_{k=0}^{N-1} c_k \phi(2t - 2m - k) \sum_{l=0}^{N-1} c_l \phi(2t - 2n - l)dt \\ &= \int_{-\infty}^{\infty} \sum_{k=0}^{N-1} c_k \sum_{l=0}^{N-1} c_l \phi(2t - 2m - k)\phi(2t - 2n - l)dt \end{aligned}$$

Substitute $2t - k = y$. Hence, $dt = \frac{1}{2}dy$ and $2t = y + k$. Thus the equation can be modified as

$$
\begin{aligned}
\int_{-\infty}^{\infty} \phi(t-m)\phi(t-n)dt &= \frac{1}{2}\int_{-\infty}^{\infty} \sum_{k=0}^{N-1} c_k \sum_{l=0}^{N-1} c_l \phi(y)\phi(y+k-2n-l)dy \\
&= \frac{1}{2}\sum_{k=0}^{N-1} c_k \sum_{l=0}^{N-1} c_l \delta_{2m,2n+l-k} \\
&= \frac{1}{2}\sum_{k=0}^{N-1} c_k \sum_{k=0}^{N-1} c_{k+2m-2n} \\
&= \frac{1}{2}\sum_{k=0}^{N-1} c_k c_{k-2z} \qquad \text{substituting } n-m \text{ as } z \\
&= \frac{1}{2}\sum_{k=0}^{N-1} c_k c_{k+2z}
\end{aligned}
$$

The same can be written in terms of filter bank coefficient as

$$\sum_{k=0}^{N-1} h_k h_{k-2z} = \delta_{z,0} \tag{5.41}$$

When $z = 0$, we have the square normalization condition, i.e.,

$$\boxed{\sum_{k=0}^{N-1} h_k^2 = 1} \tag{5.42}$$

When $z \neq 0$, we have the condition for orthogonality of translates of scaling function called **double shift orthogonality**. This condition in turn implies that the length of scaling function coefficients should be even. Assume the converse, that is, the length is odd. For example, let the length be 3. Then shifting the coefficients by 2 and taking the dot product, we get $h_2 h_0 = 0$, which is not possible unless one of the coefficients is equal to zero (refer Fig. 5.14). So the number of coefficients should be even.

Property 3: Orthogonality of Scaling Function and Wavelet Function
The scaling function and the wavelet function are orthogonal to each other. So we can write $\int \phi(t)\psi(t) = 0$. On substituting the values from refinement relations and

Fig. 5.14 Even shifting condition when the number of coefficients is odd

h_0	h_1	h_2		
		h_0	h_1	h_2

going through a series of derivation steps (omitted here), we get the final result as

$$\boxed{g_k = (-1)^k h_{N-k-1}} \tag{5.43}$$

This equation implies that, once we know the scaling function coefficients, in order to obtain the wavelet function coefficients, just reverse the scaling coefficients and change the sign of alternate values.

Property 4: Vanishing Moment Condition

In order to understand this, first we would know what are moments. The moment of order m, denoted as M_m, of a function $f(x)$ in the interval (a, b) is given by

$$M_m = \int_a^b x^m f(x)dx \tag{5.44}$$

Vanishing moment means that moment of a particular order vanishes. In the case of wavelets, the function $f(x)$ is our wavelet function $\psi(t)$. So here the pth moment of wavelet function is given by

$$M_p = \int_a^b t^p \psi(t)dt \tag{5.45}$$

We are taking moments of our wavelet function. This vanishing moment condition is a measure of quality of the wavelet. We will see how. Consider the case of Haar wavelet. The zeroth moment is given by

$$\begin{aligned} M_0 &= \int_0^1 t^0 \psi(t)dt \\ &= \int_0^{\frac{1}{2}} 1dt + \int_{\frac{1}{2}}^1 (-1)dt \\ &= 0 \end{aligned} \tag{5.46}$$

So the zeroth moment of Haar vanishes. Now take the first moment.

$$\begin{aligned} M_1 &= \int_0^1 t^1 \psi(t)dt \\ &= \int_0^{\frac{1}{2}} tdt + \int_{\frac{1}{2}}^1 (-t)dt \\ &= \frac{-1}{4} \\ &\neq 0 \end{aligned} \tag{5.47}$$

That is, the first moment of Haar does not vanish. The higher the order of moment that vanishes, the higher is the quality of the wavelet. Because the vanishing moment is an indication of the ability of the wavelet to identify the discontinuities present in the underlying signal. In other words, it indicates whether the wavelet can detect higher order derivatives or not. Haar mother wavelet has only one moment that is vanishing, i.e., the zeroth moment. So it can detect only first-order derivatives. That is why in the beginning of the chapter, we had seen that when a constant signal is given as input to a Haar filter bank, the output of the high-pass filter is zero. This is the disadvantage of Haar wavelets too. It cannot detect higher order derivatives.

When we started the design steps, we mainly concentrated on scaling function coefficients. But when it comes to the case of vanishing moments, we are talking about the wavelet function. Why? The answer is that a discontinuity indicates the presence of high frequency. So we should concentrate on the high-pass filter section of the filter bank. The high-pass section is related to the wavelet function. But in the case of denoising applications, we will go for vanishing moment conditions of scaling function too.

We know the wavelet function coefficients are related to the scaling function coefficients (Eq. 5.43). So we can convert the vanishing moment condition in terms of scaling function as: If we want the pth moment of a wavelet to vanish, then put

$$\int t^p \psi(t)dt = 0 \Rightarrow \boxed{\sum_k (-1)^k k^p h_k = 0} \tag{5.48}$$

Properties 5 and 6 discussed next are redundant. But we will discuss the equations only.

Property 5: Orthonormality of Wavelet Function to Its Translates
We know the wavelet function $\psi(t)$ is orthogonal to its own translates. It can be written in terms of filter coefficients as

$$\sum_{k=0}^{N-1} g(k)g(k+2l) = 0 \qquad for\ all\ l \neq 0 \tag{5.49}$$

Property 6: Orthonormality of Scaling Function to Wavelet Function Translates
The scaling function $\phi(t)$ is orthonormal to the translates of the wavelet function $\psi(t)$. This can be written in terms of filter coefficients as

$$\sum_{k=0}^{N-1} h(k)g(k+2l) = 0 \qquad for\ all\ l \neq 0 \tag{5.50}$$

While designing wavelets, we use only those properties that are required.

Design of Daubechies 2 Wavelet Using Equations

Daubechies 2 wavelet has filters of length 4. So $N = 4$. We can form equations based on the above properties. The four scaling function coefficients are named as h_0, h_1, h_2, h_3 (normalized coefficients).

$$\begin{aligned} h_0 + h_1 + h_2 + h_3 &= \sqrt{2} \quad \text{(normalization property)} \\ h_0^2 + h_1^2 + h_2^2 + h_3^2 &= 1 \quad \text{(square normalization property)} \\ h_0h_2 + h_1h_3 &= 0 \quad \text{(double shift orthogonality)} \\ h_0 - h_1 + h_2 - h_3 &= 0 \quad \text{(vanishing moment condition, p=0)} \\ 0.h_0 - 1.h_1 + 2.h_2 - 3.h_3 &= 0 \quad \text{(vanishing moment condition, p=1)} \end{aligned} \tag{5.51}$$

Add the equations corresponding to normalization property and vanishing moment condition, p =0.

$$\begin{aligned} 2h_0 + 2h_2 &= \sqrt{2} \\ 2(h_0 + h_2) &= \sqrt{2} \\ h_0 + h_2 &= \frac{1}{\sqrt{2}} \end{aligned}$$

Squaring, we get

$$h_0^2 + h_2^2 + 2h_0h_2 = \frac{1}{2}$$

Now subtract the equations corresponding to normalization property and vanishing moment condition, p =0.

$$\begin{aligned} 2h_1 + 2h_3 &= \sqrt{2} \\ 2(h_1 + h_3) &= \sqrt{2} \\ h_1 + h_3 &= \frac{1}{\sqrt{2}} \end{aligned}$$

Squaring, we get

$$h_1^2 + h_3^2 + 2h_1h_3 = \frac{1}{2}$$

Adding the equation now obtained

$$h_0^2 + h_2^2 + h_1^2 + h_3^2 + 2h_0h_2 + 2h_1h_3 = 1$$
$$h_0^2 + h_2^2 + h_1^2 + h_3^2 + 2(h_0h_2 + h_1h_3) = 1$$

We know from double shift orthogonality property that the term in bracket is zero.

$$h_0^2 + h_2^2 + h_1^2 + h_3^2 = 1$$

Hence, we can obtain the equation corresponding to square normalization from the equations of orthogonality, double shift orthogonality, and vanishing moment condition with p =0. So effectively, we have four equations and four unknowns. Solving these equations, the values obtained are

$$\begin{aligned} h_0 &= 0.4830, \quad h_1 = 0.8365, \quad h_2 = 0.2241 \quad \text{and} \quad h_3 = -0.1294 \\ c_0 &= 0.6830, \quad c_1 = 1.1830, \quad c_2 = 0.3170 \quad \text{and} \quad c_3 = -0.1830 \end{aligned} \tag{5.52}$$

These are same as the values obtained using the previous method. Using Eq. 5.43, the values for wavelet function coefficients are obtained as

$$\begin{aligned} g_0 &= 0.1294, \quad g_1 = 0.2241, \quad g_2 = -0.8365 \quad \text{and} \quad g_3 = 0.4830 \\ d_0 &= 0.1830, \quad d_1 = 0.3170, \quad d_2 = -1.1830 \quad \text{and} \quad d_3 = 0.6830 \end{aligned} \tag{5.53}$$

5.6 Coiflet Wavelets

Coiflets are discrete wavelets which are also designed by Ingrid Daubechies, at the request of Ronald Coifman. In the design of Coiflet wavelet, we impose vanishing moment condition on scaling function too. The vanishing moments of $\phi(t)$ make the scaling function more symmetrical. If the length of filter is equal to $6p$, then a $2p$ number of vanishing moment conditions are imposed on wavelet function, $2p - 1$ on scaling function, and the remaining on normality and orthogonality conditions, i.e., if the length of the filter is N, $\frac{N}{3}$ vanishing moment conditions are imposed on wavelet function and $\frac{N}{3} - 1$ on scaling function. Other features of Coiflet wavelets are orthogonality and compact support.

Here we will take the example of Coiflet 1 wavelet which has a length equal to 6 (minimum is 6). So here $p = 1$. We should have $2p = 2$ vanishing moment conditions on wavelet function and $2p - 1 = 1$ vanishing moment condition on scaling function. In addition, we have other properties like normalization, orthogonality, etc. The vanishing moment conditions on scaling function are new to us. It imposes condition on scaling function coefficients as

$$\begin{aligned} &\int t^n \phi(t)dt = 0 \quad \text{for } n = 1, 2 \ldots\ldots .2p-2 \\ &\Rightarrow \sum_{k=0}^{N-1} k^p h_k = \alpha^p \end{aligned} \tag{5.54}$$

where α is a real number. So we have the following equations.

$$\begin{aligned} &h_0 + h_1 + h_2 + h_3 + h_4 + h_5 = \sqrt{2} \\ &\qquad \text{(normalization property)} \\ &h_0^2 + h_1^2 + h_2^2 + h_3^2 + h_4^2 + h_5^2 = 1 \\ &\qquad \text{(square normalization property)} \\ &h_0 h_2 + h_1 h_3 + h_2 h_4 + h_3 h_5 = 0 \\ &\qquad \text{(double shift orthogonality)} \\ &h_0 h_4 + h_1 h_5 = 0 \\ &\qquad \text{(orthogonality of } \phi(t) \text{ to its translates)} \\ &h_0 - h_1 + h_2 - h_3 + h_4 - h_5 = 0 \\ &\qquad \text{(vanishing moment condition on} \psi(t) \text{, p=0)} \\ &0.h_0 - 1.h_1 + 2.h_2 - 3.h_3 + 4.h_4 - 5.h_5 = 0 \\ &\qquad \text{(vanishing moment condition on } \psi(t) \text{, p=1)} \\ &0.h_0 + 1.h_1 + 2.h_2 + 3.h_3 + 4.h_4 + 5.h_5 = \alpha \\ &\qquad \text{(vanishing moment condition on } \phi(t) \text{, p=0)} \end{aligned} \tag{5.55}$$

Solving these equations, the values for the coefficients are obtained as

$$\begin{aligned} &h_0 = -0.0727, \; h_1 = 0.3379, \; h_2 = 0.8526, \; h_3 = 0.3848, \\ &\qquad h_4 = -0.0727, \; h_5 = -0.0156 \\ &c_0 = -0.1028, \; c_1 = 0.4778, \; c_2 = 1.2058, \; c_3 = 0.5442, \\ &\qquad c_4 = -0.1028, \; c_5 = -0.0221 \end{aligned} \tag{5.56}$$

The corresponding wavelet function coefficients are easy to find. The scaling and wavelet functions are plotted in Fig. 5.15.

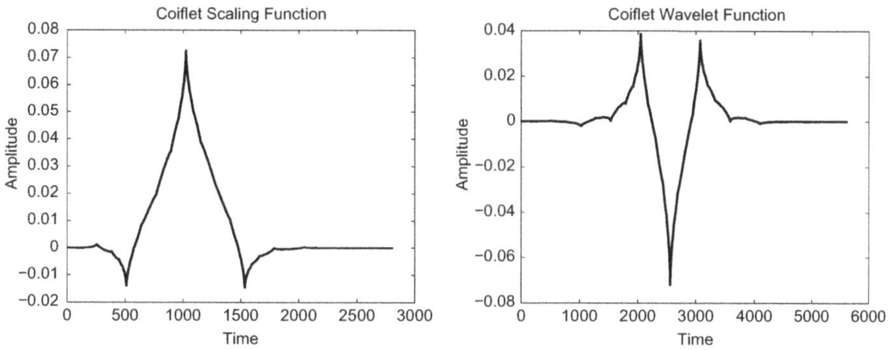

(a) Coiflet 1 scaling function $\phi(t)$

(b) Coiflet 1 wavelet function $\psi(t)$

Fig. 5.15 Coiflet 1 scaling and wavelet functions

5.7 Other Common Wavelets

Now we can have a look at some other useful wavelets.

Morlet

This is used in continuous wavelet transform. It is a complex wavelet. This wavelet has no scaling function. The wavelet function is given by

$$\psi(t) = \frac{1}{\sqrt{2\pi}} e^{-j\omega_0 t} e^{-t^2/2} \tag{5.57}$$

where ω_0 is the central frequency of the wavelet. The corresponding frequency domain equation is

$$\psi(\omega) = e^{-(w-w_0)^2}/2 \tag{5.58}$$

Here we define a new term, i.e., the admissibility condition. A wavelet is said to be admissible if the Fourier transform of the wavelet function $\psi(t)$ satisfies the condition

$$C_\psi = \int_0^\infty \frac{|\ \psi(f)\ |^2}{|\ f\ |} df < \infty \tag{5.59}$$

The Morlet wavelet is not admissible as $\psi(\omega) \neq 0$ at $\omega = 0$. But its value is very small at $\omega = 0$ and hence does not cause much problems when used in practical applications. The wavelet function of Morlet wavelet is shown in Fig. 5.16.

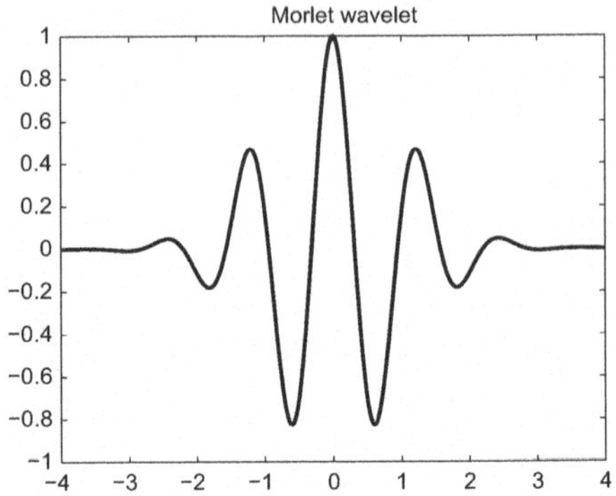

Fig. 5.16 Morlet wavelet function

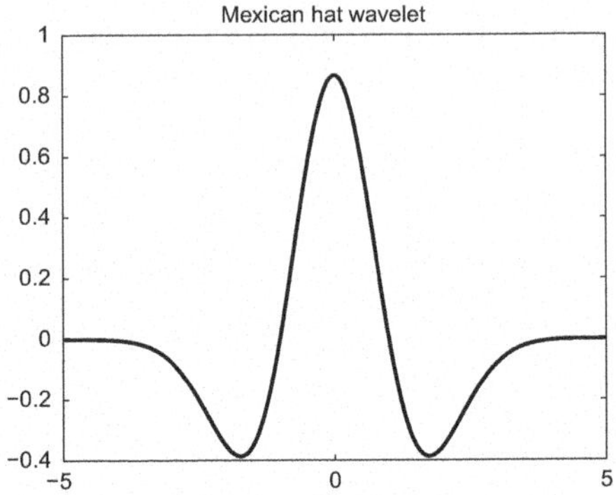

Fig. 5.17 Mexican hat wavelet function

This wavelet is mainly used for the analysis of ECG signals.

Mexican Hat Wavelet

This wavelet is also used in continuous wavelet transform. This is the second derivative of Gaussian function. Any derivative of Gaussian function can be used as a mother wavelet. If p^{th} derivative is used, it has p vanishing moments. Mexican hat wavelet has two vanishing moments, and it satisfies admissibility condition. This has also no scaling function, and the wavelet function is given by

$$\psi(t) = \frac{2}{\sqrt{3}\pi^{\frac{1}{4}}}(1 - t^2)e^{-t^2/2} \tag{5.60}$$

The wavelet function is shown in Fig. 5.17.

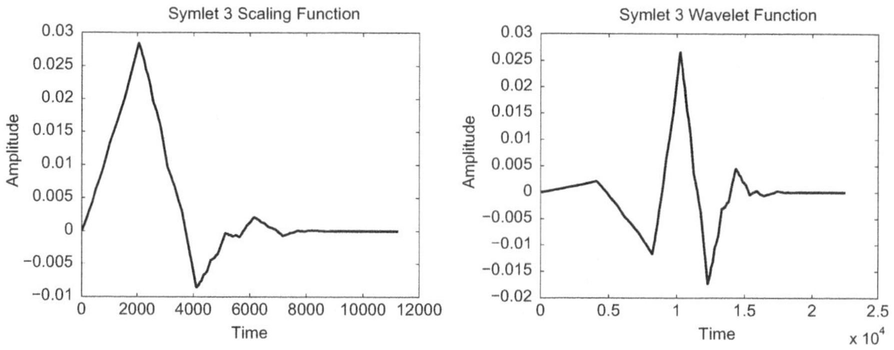

(a) Symlet 3 scaling function $\phi(t)$ (b) Symlet 3 wavelet function $\psi(t)$

Fig. 5.18 Symlet 3 scaling and wavelet functions

Symlets

In the design of Daubechies wavelet, we had seen that she had taken the roots that are inside the unit circle to get the minimum phase condition. There are other options too. They lead to symmetric wavelets. Symlets are orthogonal and has compact support. The length of the filter is $2p$ and has p vanishing moments. The properties of Daubechies family of wavelets and Symlets are almost same. The scaling and wavelet functions for Symlet 3 are shown in Fig. 5.18.

5.8 Solved Problems

Problem Draw the frequency response of Daubechies 2 wavelet. Compare with that of Haar. What inference can be obtained from this?

Solution In Daubechies 2 wavelet, the transfer function for the low-pass filter is of the form $(1 + z^{-1})^2(1 + B_0z^{-1})$ (Fig. 5.19). Simplifying it, we get

$$| H(w) |= (2cos(\omega/2))^2 Re(w) \tag{5.61}$$

On plotting, we will obtain it as shown in Fig. 5.19a and phase as in Fig. 5.19b.

So let us plot the magnitude responses for different higher order wavelets . They are shown in Fig. 5.20. We can see that as the order increases, the filter responses will become closer and closer to the ideal response.

We had plotted for the low-pass filter on the analysis side. The high-pass filter response can be easily derived from this.

Problem Derive the equations for scaling function in the calculation of Daubechies 3 wavelet from the basic Haar filter bank.

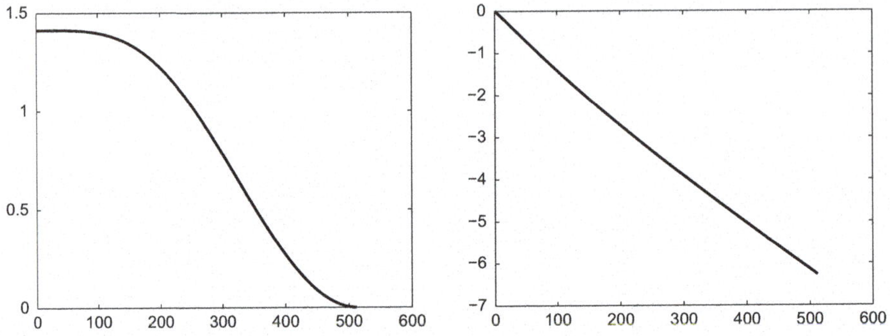

(a) Daubechies 2 magnitude response (b) Daubechies 2 phase response

Fig. 5.19 Daubechies 2 frequency response

Fig. 5.20 Magnitude response for different wavelets

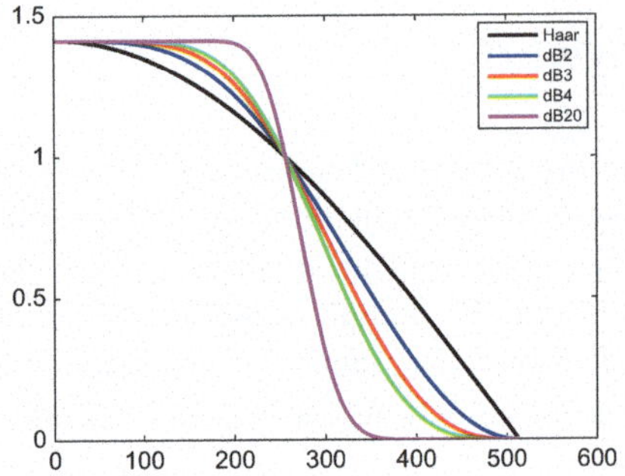

Solution Daubechies 3 wavelet has six coefficients, i.e., h_0, h_1, h_2, h_3, h_4, and h_5. This has three $(1+z^{-1})$ terms in the low-pass filter and correspondingly three $(1-z^{-1})$ terms in the high-pass filter. Now we got three zeros. We need two more zeros. Let them be $\acute{B}_0$ and $\acute{B}_1$. Now our transfer function is of the form $H(z) = C_1(1+z^{-1})^3(1+\acute{B}_0 z^{-1})(1+\acute{B}_1 z^{-1})$. Expanding, we get

$$\begin{aligned} h_0 + h_1 z^{-1} + h_2 z^{-2} + h_3 z^{-3} + h_4 z^{-4} + h_5 z^{-5} \\ &= C_1(1+3z^{-1}+3z^{-2}+z^{-3})(1+\acute{B}_0 z^{-1})(1+\acute{B}_1 z^{-1}) \\ &= 1 + (\acute{B}_0 + \acute{B}_1 + 3)z^{-1} + (3\acute{B}_0 + 3\acute{B}_1 + \acute{B}_0\acute{B}_1 + 3)z^{-2} \\ &\quad + (3\acute{B}_0 + 3\acute{B}_1 + 3\acute{B}_0\acute{B}_1 + 1)z^{-3} + (\acute{B}_0 + \acute{B}_1 + 3\acute{B}_0\acute{B}_1)z^{-4} + \acute{B}_0\acute{B}_1 z^{-5} \end{aligned} \tag{5.62}$$

Thus we get

$$
\begin{aligned}
h_0 &= 1 \\
h_1 &= \acute{B}_0 + \acute{B}_1 + 3 \\
h_2 &= 3\acute{B}_0 + 3\acute{B}_1 + \acute{B}_0\acute{B}_1 + 3 \\
h_3 &= 3\acute{B}_0 + 3\acute{B}_1 + 3\acute{B}_0\acute{B}_1 + 1 \\
h_4 &= \acute{B}_0 + \acute{B}_1 + 3\acute{B}_0\acute{B}_1 \\
h_5 &= \acute{B}_0\acute{B}_1
\end{aligned} \tag{5.63}
$$

Now consider the orthogonality conditions. As mentioned earlier, C_1 does not affect orthogonality. Consider even shift of impulse response coefficients (Fig. 5.21). So we get two equations.

$$
\begin{aligned}
h_0h_2 + h_1h_3 + h_2h_4 + h_3h_5 &= 0 \\
h_0h_4 + h_1h_5 &= 0
\end{aligned} \tag{5.64}
$$

Substituting the values in the above equation and solving them, we get the values. The impulse response coefficients are

$$
\begin{aligned}
h_0 &= 0.3327 \\
h_1 &= 0.8069 \\
h_2 &= 0.4599 \\
h_3 &= -0.1350 \\
h_4 &= -0.0854 \\
h_5 &= 0.0352
\end{aligned} \tag{5.65}
$$

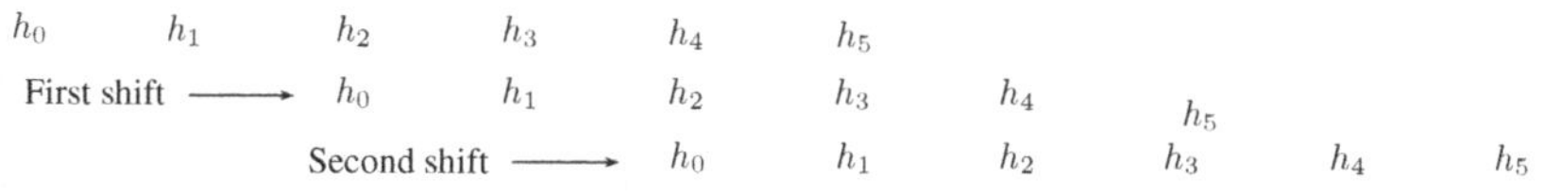

Fig. 5.21 Figure showing even shifts of $H_0[z]$

Chapter 6
Biorthogonal Wavelets

Till now we had discussed only about orthogonal filters. The orthogonal filter bank structure employs the same filters ($h[n]$ and $g[n]$) in both analysis and synthesis sides. This indirectly means that same filters are used for both decomposition and reconstruction. But the filter coefficients are not symmetrical there. In order to achieve linear phase, filters with symmetrical coefficients are required. In the case of orthogonal wavelets,indexOrthogonal wavelets only Haar wavelet is having symmetrical coefficients. But as we had seen in the last chapter, Haar is inadequate for many applications. Then there should be another way to have symmetrical filter coefficients which lead us to the concept of biorthogonality. In the case of biorthogonal wavelets, in the filter bank structure, the analysis and synthesis filters are different, and even their length may not be the same on two sides. In order to understand the idea better, let us start from the term biorthogonal itself.

6.1 Signal Space Representation of Biorthogonal Wavelets

Consider two orthogonal unit vectors $\hat{x_1}$ and $\hat{x_2}$ as shown in Fig. 6.1a. They are not collinear vectors but are perpendicular to each other and hence form the orthogonal basis for $\mathcal{R}^2$, i.e., any vector $\mathbf{A}$ can be represented as $\mathbf{A} = p_1\hat{x_1} + p_2\hat{x_2}$. The values p_1 and p_2 are obtained by projecting or taking dot product of $\mathbf{A}$ with $\hat{x_1}$ and $\hat{x_2}$, respectively (Fig. 6.1b). Now consider two other unit vectors $\hat{\hat{x_1}}$ and $\hat{\hat{x_2}}$ which are not orthogonal (Fig. 6.1c). Certainly, they form a nonorthogonal basis for $\mathcal{R}^2$. Now represent the same $\mathbf{A}$ in this nonorthogonal space. So our vector $\mathbf{A}$ can be written as $\mathbf{A} = q_1\hat{\hat{x_1}} + q_2\hat{\hat{x_2}}$ (Fig. 6.1d).

How can we find the coefficients q_1 and q_2? We cannot find q_1 and q_2 by taking dot product with $\hat{\hat{x_1}}$ and $\hat{\hat{x_2}}$ as they are not orthogonal. Now consider two other vectors $\hat{\hat{y_1}}$ and $\hat{\hat{y_2}}$ such that $\hat{\hat{x_1}} \perp \hat{\hat{y_2}}$ and $\hat{\hat{x_2}} \perp \hat{\hat{y_1}}$. Now if we take $< \mathbf{A}, \hat{\hat{y_1}} >$, we get q_1 because the component of $\mathbf{A}$ in the direction of $\hat{\hat{x_2}}$ becomes zero due to the

M. S. Sinith et al., *Wavelets and Multi-Resolution Analysis*,
https://doi.org/10.1007/978-3-031-92750-8_6

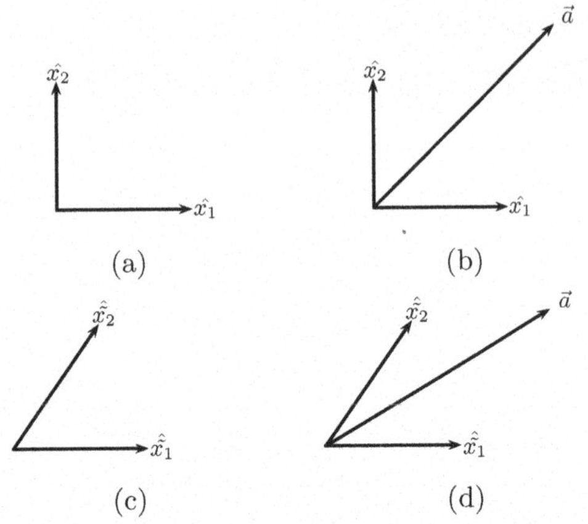

Fig. 6.1 Orthogonal and nonorthogonal vectors that form the bases of $\mathcal{R}^2$

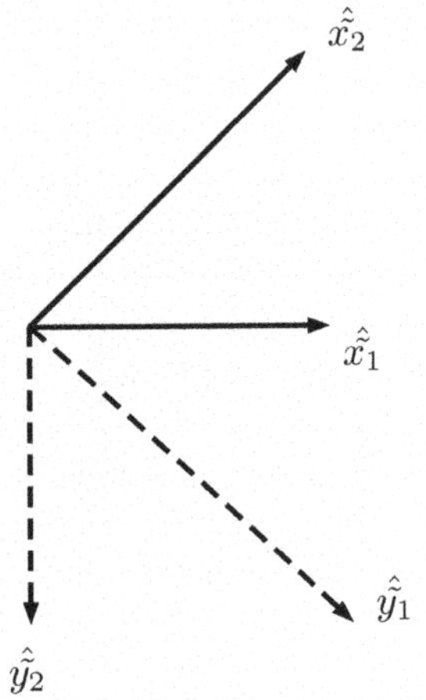

Fig. 6.2 Figure showing biorthogonal vectors

property that $\hat{\tilde{y}}_1 \perp \hat{\tilde{x}}_2$. Similarly to find q_2, take $< \mathbf{A}, \hat{\tilde{y}}_2 >$. Thus $\hat{\tilde{y}}_1$ and $\hat{\tilde{y}}_2$ form a biorthogonal basis of $\hat{\tilde{x}}_1$ and $\hat{\tilde{x}}_2$ (Fig. 6.2).

Let us take an example and see how biorthogonal vectors are formed.

Consider the vectors $\hat{\tilde{x}}_1 = (1, 0)$ and $\hat{\tilde{x}}_2 = (2, 5)$. They are not orthogonal among themselves. Consider a matrix T with rows as our vectors $\hat{\tilde{x}}_1$ and $\hat{\tilde{x}}_2$.

$$T = \begin{bmatrix} 1 & 0 \\ 2 & 5 \end{bmatrix}$$

Now find the inverse of T.

$$T^{-1} = \begin{bmatrix} 1 & 0 \\ \frac{-2}{5} & \frac{1}{5} \end{bmatrix}$$

Form two vectors $\hat{\hat{y}}_1$ and $\hat{\hat{y}}_2$ which are the columns of T^{-1}. So $\hat{\hat{y}}_1 = (1, \frac{-2}{5})$ and $\hat{\hat{y}}_2 = (0, \frac{1}{5})$. $\hat{\hat{y}}_1$ and $\hat{\hat{y}}_2$ are the dual of $\hat{\hat{x}}_1$ and $\hat{\hat{x}}_2$. Now we can note that $\hat{\hat{y}}_1$ and $\hat{\hat{y}}_2$ are also not orthogonal. Let $\langle \hat{\hat{x}}_1, \hat{\hat{y}}_2 \rangle$ represent the dot product of vectors $\hat{\hat{x}}_1$ and $\hat{\hat{y}}_2$. So we get the following.

$$\begin{aligned} \langle \hat{\hat{x}}_1, \hat{\hat{y}}_1 \rangle &= 1 \\ \langle \hat{\hat{x}}_1, \hat{\hat{y}}_2 \rangle &= 0 \\ \langle \hat{\hat{x}}_2, \hat{\hat{y}}_1 \rangle &= 0 \\ \langle \hat{\hat{x}}_2, \hat{\hat{y}}_2 \rangle &= 1 \end{aligned} \tag{6.1}$$

The vectors $\hat{\hat{x}}_1$ and $\hat{\hat{x}}_2$ are not orthogonal among themselves, but $\hat{\hat{x}}_1$ is orthogonal to $\hat{\hat{y}}_2$ and $\hat{\hat{x}}_2$ is orthogonal to $\hat{\hat{y}}_1$.

Thus in general, suppose we have a set of n independent vectors $\{\hat{\hat{x}}_1, \hat{\hat{x}}_2, \ldots, \hat{\hat{x}}_n\}$. These vectors are not orthogonal among themselves. Write a matrix with these vectors as the rows. Find the inverse of the matrix, and take the columns of the inverse matrix as vectors. Then we get another n vectors, let them be $\{\hat{\hat{y}}_1, \hat{\hat{y}}_2, \ldots, \hat{\hat{y}}_n\}$. They are related as

$$\begin{aligned} \hat{\hat{x}}_i \perp \hat{\hat{y}}_j &\text{ for every } i \neq j \\ \langle \hat{\hat{x}}_i, \hat{\hat{y}}_j \rangle = 0 &\text{ for every } i \neq j \\ \langle \hat{\hat{x}}_i, \hat{\hat{y}}_i \rangle = 1 &\text{ for every } i \end{aligned} \tag{6.2}$$

A function $f(t)$ in vector space V_j can be written as

$$f(t) = \sum_k s_j(k) 2^{\frac{j}{2}} \phi(2^j t - k) \tag{6.3}$$

In order to find $s_j(k)$, if it is the orthogonal system, we just project $s_j(k)$ on to $\phi_{j,k}(t)$. But here we have to project to the dual basis $\tilde{\phi}_{j,k}(t)$.

Therefore,

$$s_j(k) = \int f(t) \tilde{\phi}_{j,k}(t) dt \tag{6.4}$$

6.2 Classical Wavelets and Biorthogonal Wavelets

The wavelets which are translations and dilations of a single function are called classical wavelets. There will be only one scaling function and one wavelet function.

In other words, the analysis and synthesis filters are the same. Usually, the filters are not symmetrical, which means that we will not have linear phase. The need for biorthogonal wavelets is that we want linear phase by making the synthesis filters different from analysis filters. Also the design of biorthogonal filters gives us more degrees of freedom so that the linear phase condition can be incorporated. Thus we can achieve linear phase, and also the reconstruction will be much better with the use of biorthogonal wavelets.

As we have only one low-pass and high-pass filter in orthogonal filter bank, there is only one scaling and wavelet function. But in biorthogonal case, there will be two scaling functions. They are called as primary scaling function and dual scaling function. In the same way, we have two wavelet functions, called primary wavelet function and dual wavelet function. If the primary scaling function is denoted by $\phi(t)$, the dual scaling function is indicated as $\tilde{\phi}(t)$. Similarly for wavelet function, primary and dual are represented as $\psi(t)$ and $\tilde{\psi}(t)$, respectively. From refinement relations, we know

$$\phi(t) = \sum_{k=0}^{N-1} h_k \sqrt{2}\phi(2t-k) \tag{6.5}$$

Similarly, the dual scaling function will also satisfy this relation.

$$\tilde{\phi}(t) = \sum_{k=0}^{N-1} \tilde{h}_k \sqrt{2}\tilde{\phi}(2t-k) \tag{6.6}$$

For completion, we can have the relations for wavelet functions too.

$$\begin{aligned} \psi(t) &= \sum_{k=0}^{N-1} g_k \sqrt{2}\phi(2t-k) \\ \tilde{\psi}(t) &= \sum_{k=0}^{N-1} \tilde{g}_k \sqrt{2}\tilde{\phi}(2t-k) \end{aligned} \tag{6.7}$$

In orthogonal system, $\phi(t)$ is orthogonal to its own integer translates. In biorthogonal system, $\phi(t)$ is not orthogonal to its own integer translates but is orthogonal to integer translates of $\tilde{\phi}(t)$.

That is,

$$\begin{aligned} &\int \phi(t-k)\tilde{\phi}(t-m)dt = 0 \text{ for } k \neq m \\ &\int \phi(t-k)\tilde{\phi}(t-k)dt = 1 \text{ for every } k \\ &\text{i.e.,} \int \phi(t)\tilde{\phi}(t-k)dt = \delta_{k,0} \end{aligned} \tag{6.8}$$

Similarly, the condition that $\phi(t)$ is orthogonal to $\psi(t)$ and its translates is not true in biorthogonal case. Instead, $\phi(t)$ is orthogonal to $\tilde{\psi}(t)$ and its translates.

The wavelet function coefficients are obtained from scaling function coefficients in orthogonal case using the property that the scaling and wavelet functions are orthogonal to each other. The relation is given by

$$g_k = (-1)^k h_{N-k-1} \tag{6.9}$$

But in biorthogonal case, this relation is not valid. But the dual wavelet function is perpendicular to the primary scaling function. Hence, the dual wavelet function coefficients are given by

$$\tilde{g}_k = (-1)^k h_{N-k-1} \tag{6.10}$$

In the same way, we can relate the primary wavelet function coefficients and dual scaling function coefficients.

$$g_k = (-1)^k \tilde{h}_{N-k-1} \tag{6.11}$$

Hence,

$$\begin{aligned} \psi(t) &= \sum_{k=0}^{N-1} (-1)^k \tilde{h}_{N-k-1} \sqrt{2} \tilde{\phi}(2t-k) \\ \tilde{\psi}(t) &= \sum_{k=0}^{N-1} \tilde{(} - 1)^k h_{N-k-1} \sqrt{2} \phi(2t-k) \end{aligned} \tag{6.12}$$

As in the case of scaling function, the wavelet function $\psi(t)$is not orthogonal to its own translates, but orthogonal to the translates of $\tilde{\psi}(t)$.

$$\int \psi(t-k)\tilde{\psi}(t-m)dt = \delta_{k-m} \tag{6.13}$$

6.3 B-Spline Wavelets

B-splines form a system where the scaling function coefficients are the binomial coefficients. B-splines of zeroth order is Haar. In Haar case by using dilation equation, we can write

$$\phi_0(t) = \phi_0(2t) + \phi_0(2t-1) \tag{6.14}$$

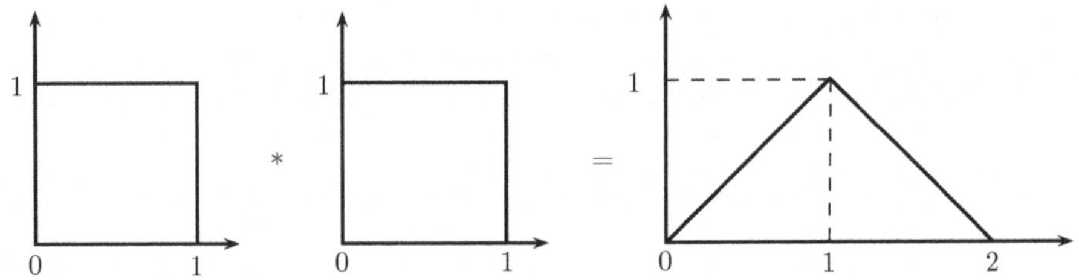

Fig. 6.3 Scaling function for second-order spline

The scaling function of spline of order 2 is obtained by convolving $\phi_0(t)$ with itself. The result is shown in Fig. 6.3.

This scaling function can be expressed as

$$\phi_1(t) = \frac{1}{2}\phi_1(2t) + \phi_1(2t-1) + \frac{1}{2}\phi_1(2t-2) \tag{6.15}$$

The coefficients of the scaling function are $\frac{1}{2}[1, 2, 1]$. $[1, 2, 1]$ are the binomial coefficients. The scaling function of spline of order 3 is obtained by convolving $\phi_0(t)$ three times. It is equivalent to convolving $\phi_1(t)$ with $\phi_0(t)$.

The scaling function is given by

$$\phi_2(t) = \frac{1}{4}\phi_2(2t) + \frac{3}{4}\phi_2(2t-1) + \frac{3}{4}\phi_2(2t-2) + \frac{1}{4}\phi_2(2t-3) \tag{6.16}$$

The coefficients of the scaling function are $\frac{1}{4}[1, 2, 1]$. $[1, 3, 3, 1]$, which are also the binomial coefficients. B-spline of second order has coefficients $[1, 2, 1]$, third order has coefficients $[1, 3, 3, 1]$, and so on.

Consider the scaling function of B-spline of order 2 shown in Fig. 6.3. It is not orthogonal to its integer translates. Specifically, $\langle\phi(t), \phi(t-1)\rangle$ and $\langle\phi(t), \phi(t+1)\rangle$ are nonzero. But all other dot products are zeros. Hence, we are going for biorthogonal filters.

6.4 B-Spline Biorthogonal Wavelet System or Cohen-Daubechies-Feauveau Wavelets

We saw that the spline wavelets of order 2 and above are not orthogonal to their integer translates. Hence, we are going to design a dual scaling function $\tilde{\phi}(t)$. First we have to decide on the length of the coefficients of $\tilde{\phi}(t)$. We know in orthogonal case the length of h should be even which is derived from the equation

$$\sum_n h(n)h(n-2k) = \delta_{k,0}$$

But in biorthogonal case, the above equation changes to

$$\sum_{n} h(n)\tilde{h}(n-2k) = \delta_{k,0} \tag{6.17}$$

which means that the difference between the lengths of h and $\tilde{h}$ should be even. Let $h(n)$ be nonzero in the interval $N_1 \le n \le N_2$ and $\tilde{h}(n)$ be nonzero in the interval $\tilde{N}_1 \le n \le \tilde{N}_2$. Then

$$\begin{aligned} N_2 - \tilde{N}_1 &= 2k+1 \text{ and} \\ \tilde{N}_2 - N_1 &= 2\tilde{k}+1 \text{ where } k \text{ and } \tilde{k} \text{ are integers} \end{aligned} \tag{6.18}$$

Let us take an example for the design of biorthogonal B-spline wavelets. Consider B-spline wavelet of order 2 whose coefficients are given by $\frac{1}{2}[1, 2, 1]$. These are not normalized coefficients. To normalize them, we know the sum should be $\sqrt{2}$.

$$c\left(\frac{1}{2}+1+\frac{1}{2}\right) = \sqrt{2} \implies c = \frac{1}{\sqrt{2}} \tag{6.19}$$

Hence, the normalized scaling function coefficients of second-order B-spline system is given by $\frac{1}{2\sqrt{2}}, \frac{1}{\sqrt{2}}, \frac{1}{2\sqrt{2}}$. The scaling function has three coefficients. Now let us take the length of dual scaling function as 5. It is not compulsory to choose 5, and we can take any positive odd number so that the difference in lengths of h and $\tilde{h}$ be even. Also we have to position the coefficients so as to satisfy Eq. 6.17. For our case, let us take $N_1 = 1, N_2 = 3, \tilde{N}_1 = 0, \tilde{N}_2 = 4$ so that $k = \tilde{k} = 1$. The positioning of coefficients is very important. If Eq. 6.17 is not satisfied, we may not get correct solution for $\tilde{h}(k)$. Now in order to achieve the linear phase property, the coefficients of dual scaling function must be symmetric. So $h(0) = h(4)$ and $h(1) = h(3)$. Now the coefficients of h and $\tilde{h}$ are as shown in Fig. 6.4. We know $\tilde{\phi}(t) \perp \psi(t)$. So

$$\psi(t) = \sum_{k=0}^{N-1} (-1)^k \tilde{h}(N-k-1)\tilde{\phi}(2t-k) \tag{6.20}$$

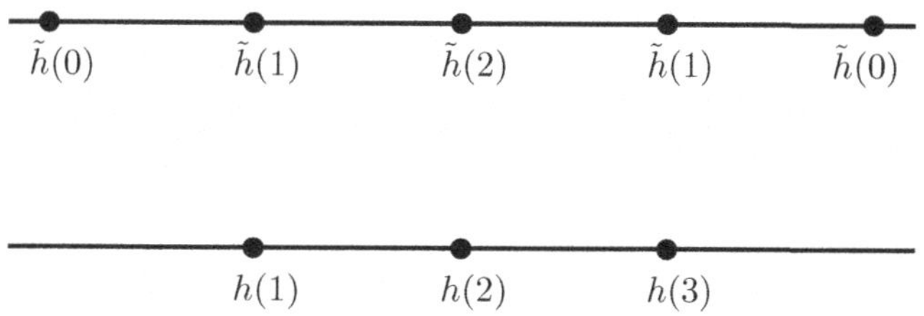

Fig. 6.4 Positioning of scaling function coefficients of second-order spline

But it is required that $\int \psi(t)dt = 0$. So

$$\sum_{k=0}^{N-1}(-1)^k \tilde{h}(N-k-1) = 0 \tag{6.21}$$

This is the vanishing moment condition.

Now we can apply conditions on coefficients:

1. **Normality**
The first condition is $\sum_k \tilde{h}(k) = \sqrt{2}$ so that $\int \tilde{\phi}(t)dt = 1$. So the first condition is

$$2\tilde{h}(0) + 2\tilde{h}(1) + \tilde{h}(2) = \sqrt{2} \tag{6.22}$$

2. **Biorthogonality**
Substitute $k = 0$ in Eq. 6.17. So we get

$$\begin{aligned} h(1)\tilde{h}(1) + h(2)\tilde{h}(2) + h(3)\tilde{h}(1) &= 1 \\ \frac{1}{2\sqrt{2}}\tilde{h}(1) + \frac{1}{\sqrt{2}}\tilde{h}(2) + \frac{1}{2\sqrt{2}}\tilde{h}(1) &= 1 \\ \text{i.e.,} \quad \frac{1}{\sqrt{2}}\tilde{h}(1) + \tilde{h}(2) &= 1 \end{aligned} \tag{6.23}$$

For $k = 1$, we get

$$\begin{aligned} h(1)\tilde{h}(1) + h(2)\tilde{h}(0) &= 0 \\ \frac{1}{2\sqrt{2}}\tilde{h}(1) + \frac{1}{\sqrt{2}}\tilde{h}(0) &= 0 \\ \text{i.e.,} \quad \frac{1}{2}\tilde{h}(1) + \tilde{h}(0) &= 0 \end{aligned} \tag{6.24}$$

For $k = -1$, we get the same set of equations for $k = 1$ because of the symmetric property of $\tilde{h}$.

3. **Vanishing moment condition**
Here the length of biorthogonal scaling function filter is 5. So $N = 5$. Hence by substituting on Eq. 6.21,

$$\begin{aligned} \tilde{h}(4) - \tilde{h}(3) + \tilde{h}(2) - \tilde{h}(1) + \tilde{h}(0) &= 0 \\ 2\tilde{h}(0) - 2\tilde{h}(1) + \tilde{h}(2) &= 0 \end{aligned} \tag{6.25}$$

Solving these four equations, we get

$$\tilde{h}(0) = \frac{-\sqrt{2}}{8}, \ \tilde{h}(1) = \frac{\sqrt{2}}{4}, \ \tilde{h}(2) = \frac{3\sqrt{2}}{4}$$

$\tilde{h}(3) = \tilde{h}(1)$ and $\tilde{h}(4) = \tilde{h}(0)$ due to symmetric property.

We got all the coefficients to generate the scaling and wavelet functions which can be summarized as below.

The scaling and wavelet functions for biorthogonal spline wavelet can be drawn as shown in Fig. 6.5.

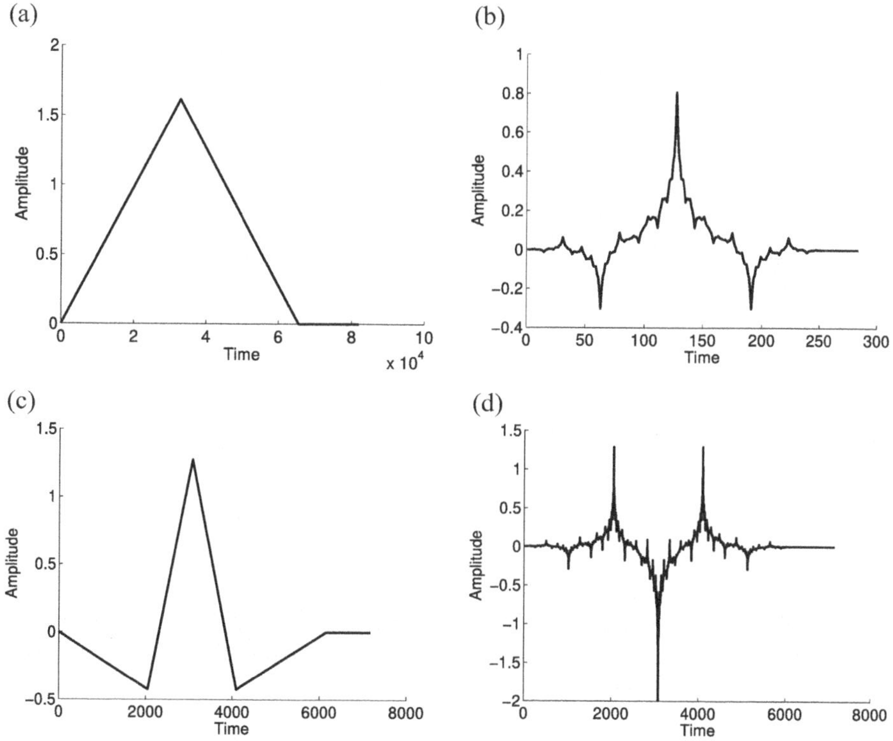

Fig. 6.5 (**a**) B Spline Scaling Function (**b**) B Spline Dual Scaling Function (**c**) B Spline Wavelet Function (**d**) B Spline Dual Wavelet Function

6.5 Lifting Scheme

This section discusses an entirely new concept in the design of wavelets. Till now we used the translations and dilations of scaling and wavelet functions to design wavelets. The wavelets so designed are called classical or first-generation wavelets. But as we saw in the beginning of this chapter, biorthogonal wavelets ae not translations and dilations of single function. The new method "lifting scheme" can be used for their construction or the so-called second-generation wavelets. Here all the concepts are in spatial domain. We are not using the conventional Fourier analysis for the design. So a person who is not having much mathematical background can understand the process better and can use wavelet transforms in different applications.

Introduction to Lifting a Simple Example

Assume that we have a sequence $x[n] = \{0, 1, 2, 3, 2, 1, 0, -1, -2, -3, -2, -1\}$. The sequence has 12 sample points. Suppose that we want to reduce the number of samples of this sequence without much loss of information. Or that we want to compress the signal. The easiest way is to downsample the signal by a factor of 2. Let us represent the downsampled signal by $x_1[n] = \{0, 2, 2, 0, -2, -2\}$. The length has now reduced to 6 as we expected, and the samples in $x_1[n]$ are the even samples of $x[n]$. Of course, we have lost something from $x[n]$. What is that? They are the odd-numbered samples. The odd samples are given by $x_2[n] = \{1, 3, 1, -1, -3, -1\}$. How can we recover the original signal $x[n]$ if we know $x1[n]$ and $x_2[n]$? We place the samples of $x1[n]$ at even time indices and in between place the samples from $x_2[n]$. Isn't it? Here in our example, we know the even and odd sets and hence recombination is easier. But practically, we will be knowing one of the two sets, and we should be able to predict the other with reasonable accuracy. What is the way to achieve this?

Suppose we are applying a simple linear transformation which replaces even and odd subsets by averages s and differences d, i.e., the sequence s can be found by taking the average of subsequent samples of $x[n]$. $s[0] = (x[0] + x[1])/2, s[1] = (x[2] + x[3])/2$ and so on. Similarly, the sequence $d[n]$ can be found by taking the difference of subsequent samples of $x[n]$. $d[0] = x[1] - x[0], d[1] = x[3] - x[2]$ and so on (Taking subsequent samples means we are taking an even sample and the immediate odd sample.). So the sequences s and d are given by

$$s = \{1/2, 5/2, 3/2, -1/2, -5/2, -3/2\}$$

$$d = \{1, 1, -1, -1, -1, 1\}$$

Can we recover our original signal from s and d? Yes of course.

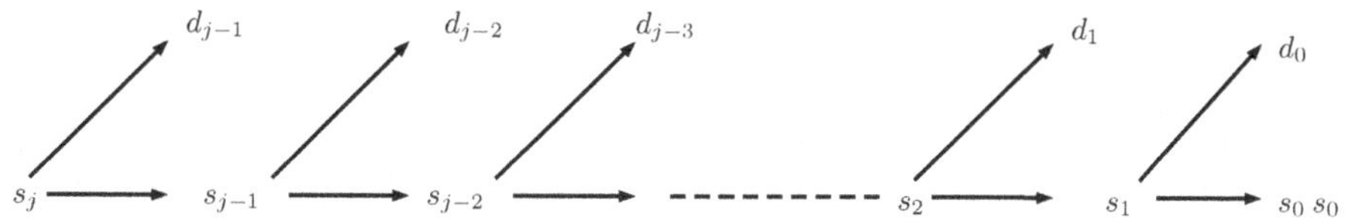

Fig. 6.6 Signal decomposition into average and difference

$$x[0] = s[0] - \frac{d[0]}{2}$$
$$x[1] = s[0] + \frac{d[0]}{2}$$
$$x[2] = s[1] - \frac{d[1]}{2}$$
$$x[3] = s[1] + \frac{d[1]}{2}$$

and so on. You can simply verify the result for our given example.

Generalizing the Concept

Now we can generalize the concept presented using the example. Consider a signal $s_{\{j\}}$ of 2^j sample values $s_{\{j,k\}}$. Applying the average and difference transform for each (even:odd) pair $s_{\{j,2k\}}$ and $s_{\{j,2k+1\}}$, we have 2^{j-1} such pairs. We are going down a level when we find the average and difference. Hence, let us denote them as $s_{\{j-1,k\}}$ and $d_{\{j-1,k\}}$.

$$\begin{aligned} s_{\{j-1,k\}} &= \frac{s_{\{j,2k\}} + s_{\{j,2k+1\}}}{2} \\ d_{\{j-1,k\}} &= s_{\{j,2k\}} - s_{\{j,2k+1\}} \end{aligned} \tag{6.26}$$

The averages $s_{\{j-1\}}$ can be thought of as a coarser representation of the signal $s_{\{j\}}$ and $d_{\{j-1\}}$ as the detail information needed to go from coarser representation back to the original signal. If our original signal has sample values that are highly correlated, the average signal will closely resemble the original signal, and the difference signal will have very small values (details will be very small).

Again we can apply the same transformation to $s_{\{j-1\}}$ and can split it into $s_{\{j-2\}}$ and $d_{\{j-2\}}$ each with 2^{j-2} samples. This can be iterated j times, i.e., till we have no further samples. The last level $s_{\{0\}}$ contains only one value $s_{\{0,0\}}$ which is the average of all the samples of the original signal $s_{\{j\}}$. It is the DC component or the zero frequency of the signal (Fig. 6.6).

Haar and Lifting

In the last section, we saw how to compute the average and difference of samples and how to recover the original signal from average and difference. Assume that we want to save the memory and want to store the average and difference in the same location of sample values, i.e., consider the first two samples of our signal $x[0]$ and $x[1]$. We compute their average and difference, let them be $s[0]$ and $d[0]$. The value $s[0]$ should be stored in the location of $x[0]$ and $d[0]$ in the location of $x[1]$. Assume that at first we computed the average $s[0]$ and stored in the location of $x[0]$. Now we have to compute the difference. If we take the difference of adjacent samples, it will be $x[1] - s[0]$, which will lead to wrong answer. To solve this problem, first compute the difference. So

$$d[0] = x[1] - x[0] \tag{6.27}$$

Then store $d[0]$ in the location of $x[1]$. Now we have to compute the difference with this new value of $d[0]$. The equation we use is

$$s[0] = x[0] + \frac{d[0]}{2} \tag{6.28}$$

Will the answer be correct?

$$s[0] = x[0] + \frac{d[0]}{2} = x[0] + \frac{x[1] - x[0]}{2} = \frac{x[0] + x[1]}{2}$$

which is our expected value of average. This value can be stored in the place of $x[0]$.

The Lifting Scheme

Consider a signal $s_{\{j\}}$ of 2^j sample values $s_{\{j,k\}}$. We want to transform this signal into a coarser signal $s_{\{j-1\}}$ and a detail signal $d_{\{j-1\}}$. The lifting scheme consists of three steps:

- Split
- Predict
- Update

The idea of lifting scheme will be more clear from Fig. 6.7.

We will see each step in detail:

1. **Split**
   This stage splits our original signal into two separate groups of signals. The first group consists of even samples $s_{\{2k\}}$, and the other group consists of odd samples $s_{\{2k+1\}}$. The number of samples in each group will be half the number of samples

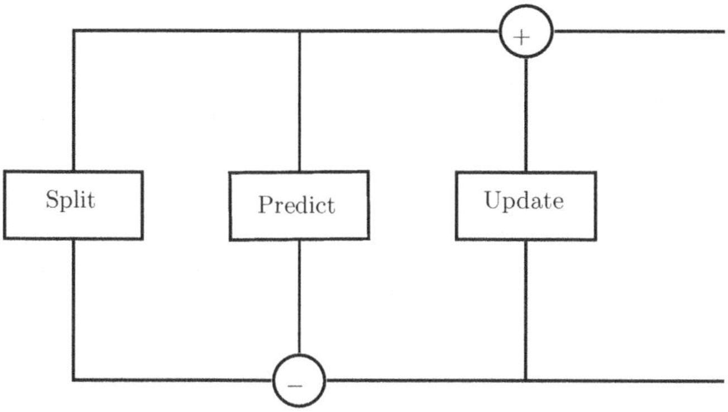

Fig. 6.7 The three steps in lifting scheme

in our original signal. This process is called lazy wavelet transform. So this stage can be explained as

$$(\text{even}_{\{j-1\}};\ \text{odd}_{\{j-1\}}) = \text{Split}(s_j)$$

2. **Predict**
Now we have two sets even and odd. If we are provided with one set, we should be able to predict the other. We always predict the odd set with the help of even set.
In Haar case, we use the adjacent left even sample to predict the next odd sample. Hence, we can predict $s_{\{j,2k+1\}}$ with the even sample $s_{\{j,2k\}}$. Then the detail coefficient $d_{\{j-1\}}$ is given by

$$d_{\{j-1,k\}} = s_{\{j,2k+1\}} - s_{\{j,2k\}}$$

Generally, we can write

$$d_{\{j-1\}} = \text{odd}_{\{j-1\}} - P(\text{even}_{\{j-1\}})$$

If the samples in original signal are highly correlated, then the detail coefficient values will be very small.
3. **Update**
The average signal $s_{\{j-1\}}$ can be found by

$$s_{\{j-1,k\}} = s_{\{j,2k\}} + \frac{d_{\{j-1,k\}}}{2}$$

We had seen earlier that the coarser signal has same average value as the original signal. The last coefficient $s_{\{0,0\}}$ will be overall average of the signal.

$$\begin{aligned}\sum_{k=0}^{2^j-1} s_{\{j-1,k\}} &= \sum_{k=0}^{2^{j-1}} s_{\{j,2k\}} + \frac{d_{\{j-1,k\}}}{2} \\ &= \frac{1}{2}\sum_{k=0}^{2^{j-1}} s_{\{j,2k\}} + s_{\{j,2k+1\}} \\ &= \frac{1}{2}\sum_{k=0}^{2^{j}} s_{\{j,l\}}\end{aligned}$$

Thus define an operator U of the form,

$$s_{\{j-1\}} = \text{even}_{\{j-1\}} + U(d_{\{j-1\}})$$

We know the even locations are replaced with averages and odd with differences. Hence, the three stages in lifting scheme can be represented in three equations as

$$\begin{aligned}(\text{even}_{\{j-1\}}; \text{odd}_{\{j-1\}}) &= \text{Split}(s_{\{j\}}) \\ \text{odd}_{\{j-1\}} &= \text{odd}_{\{j-1\}} - P(\text{even}_{\{j-1\}} \\ \text{even}_{\{j-1\}} &= \text{even}_{\{j-1\}} + U(d_{\{j-1\}})\end{aligned} \tag{6.29}$$

The Inverse Lifting Scheme

The steps in the inverse process can be explained in three steps:

- Undo update
- Undo predict
- Merge

1. **Undo Update**
   We can recover the even samples by

$$\text{even}_{\{j-1\}} = s_{\{j-1\}} - U(d_{\{j-1\}})$$

   For Haar, the same equation will be like

$$s_{\{j,2k\}} = s_{\{j-1,k\}} - \frac{d_{\{j-1,k\}}}{2}$$

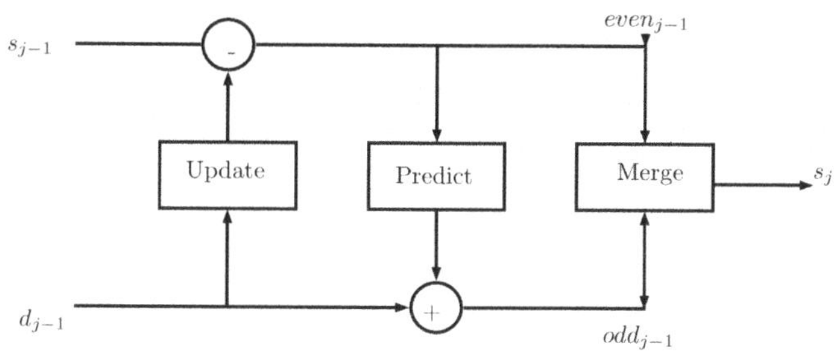

Fig. 6.8 The inverse lifting scheme

2. **Undo Predict**
   We get the even samples from undo update stage. Based on these even samples and the prediction information, we can recover the odd samples.

$$\text{odd}_{\{j-1\}} = d_{\{j-1\}} + P(\text{even}_{\{j-1\}})$$

   For Haar, the same equation will be like

$$s_{\{j,2k+1\}} = d_{\{j-1,k\}} + s_{\{j,2k\}}$$

3. **Merge**
   Now we have the even and odd samples of the signal. To recover the original signal, we simply have to merge them together. This can be thought of as the inverse of lazy wavelet transform.

$$s_{\{j\}} = \text{Merge}(\text{odd}_{\{j-1\}}, \text{even}_{\{j-1\}})$$

   The inverse lifting is shown in Fig. 6.8.

6.6 The Linear Wavelet Transform Using Lifting

The lifting scheme we saw till now consists of one predictor and one update step. Or we had seen the Haar transform in which case the original signal is a constant. If we have single predictor, we say the order of the predictor is one and same is the case with update stage. So the Haar transform has predictor and update of order one. In practical situations, we have to deal with other types of signals which are not constant. Depending on our application, we can have more number of predictor and update stages. In this section, we can see a predictor and update of order two.

Consider a signal which is linear, but not constant. An example is shown in Fig. 6.9. We use a predicting scheme in which an odd sample will be the average of immediate left neighboring even sample and immediate right neighboring even sample. So by referring Eq. 6.29, we can write $P(\text{even}_{\{j-1\}}) = \frac{1}{2}(s_{\{j,2l\}} + s_{\{j,2l+2\}})$. Hence, the detail coefficient, $d_{\{j-1\}}$, is given by

$$\begin{aligned} \text{odd}_{\{j-1\}} &= \text{odd}_{\{j-1\}} - P(\text{even}_{\{j-1\}}) \\ d_{\{j-1\}} &= s_{\{j,2l+1\}} - \frac{1}{2}(s_{\{j,2l\}} + s_{\{j,2l+2\}}) \end{aligned} \tag{6.30}$$

In the update stage, we make sure that the average of the signal is the same in every stage, i.e.,

$$\sum_l s_{\{j-1,l\}} = \frac{1}{2}\sum_l s_{\{j,l\}} \tag{6.31}$$

Let the update be of the form

$$\begin{aligned} \text{even}_{\{j-1\}} &= \text{even}_{\{j-1\}} + U(\text{odd}_{\{j-1\}}) \\ s_{\{j-1,l\}} &= s_{\{j,2l\}} + A(d_{\{j-1,l-1\}} + d_{\{j-1,l\}}) \end{aligned} \tag{6.32}$$

Next is to find the value of A. We use Eq. 6.31 for the purpose. First of all, find $\sum_l s_{\{j-1,l\}}$. Substituting from Eq. 6.32, we get

$$\sum_l s_{\{j-1,l\}} = \sum_l s_{\{j,2l\}} + \sum_l A(d_{\{j-1,l-1\}} + d_{\{j-1,l\}}) \tag{6.33}$$

Substitute Eq. 6.30 in Eq. 6.33. Then using the concept $\sum_l s_{\{j,2l\}} = \sum_l s_{\{j,2l-2\}} = \sum_l s_{\{j,2l+2\}}$ and similarly for the odd terms, we get

$$\sum_l s_{\{j-1,l\}} = (1-2A)\sum_l s_{\{j,2l\}} + 2A\sum_l s_{\{j,2l+1\}} \tag{6.34}$$

We know from Eq. 6.31, $\sum_l s_{\{j-1,l\}}$ should be equal to $\frac{1}{2}\sum_l s_{\{j,l\}}$. For that, the value of A should be $\frac{1}{4}$. Thus

$$s_{\{j-1,l\}} = s_{\{j,2l\}} + \frac{1}{4}(d_{\{j-1,l-1\}} + d_{\{j-1,l\}}) \tag{6.35}$$

Thus we had given a lift of $\frac{1}{4}$ of the sum of detail coefficients to an even-indexed signal (Fig. 6.9).

On the reverse,

$$s_{\{j,2l\}} = s_{\{j-1,l\}} - \frac{1}{4}(d_{\{j-1,l-1\}} + d_{\{j-1,l\}}) \tag{6.36}$$

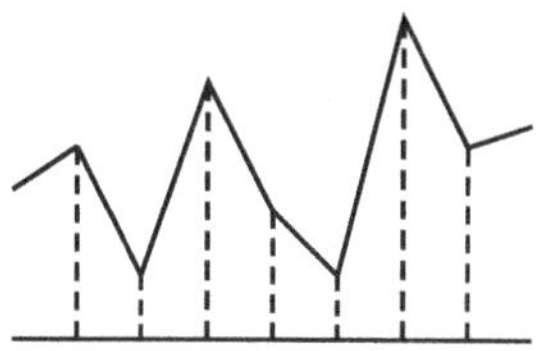

Fig. 6.9 Example for a linear function

gives us the even-indexed samples, and

$$s_{\{j,2l+1\}} = d_{\{j,l\}} - \frac{1}{2}(s_{\{j,2l\}} + s_{\{j,2l+2\}}) \tag{6.37}$$

gives us the odd indexed samples.

The wavelet transform we discussed now is the biorthogonal Cohen-Daubechies-Feauveau.

$$\begin{aligned} s_{\{j-1,l\}} &= s_{\{j,2l\}} + \frac{1}{4}(d_{\{j-1,l-1\}} + d_{\{j-1,l\}}) \\ &= s_{\{j,2l\}} + \frac{1}{4}\left(s_{\{j,2l-1\}} - \frac{1}{2}[s_{\{j,2l-2\}} + s_{\{j,2l\}}] + s_{\{j,2l+1\}}\right. \\ &\quad \left. -\frac{1}{2}[s_{\{j,2l\}} + s_{\{j,2l+2\}}]\right) \\ &= \frac{-1}{8}s_{\{j,2l-2\}} + \frac{1}{4}s_{\{j,2l-1\}} + \frac{3}{4}s_{\{j,2l\}} + \frac{1}{4}s_{\{j,2l+1\}} + \frac{-1}{8}s_{\{j,2l+2\}} \end{aligned} \tag{6.38}$$

Thus we get the dual scaling function coefficients as $\frac{-1}{8}, \frac{1}{4}, \frac{3}{4}, \frac{1}{4}, \frac{-1}{8}$, which will lead us to the same coefficients we obtained in Sect. 6.4 on normalization.

6.7 Solved Problems

Problem Find a set of biorthogonal vectors for the vectors $x_1 = (3, 1)$ and $x_2 = (\frac{1}{2}, \frac{3}{4})$ using Matlab and verify the result.

Solution The MATLAB program is given below.

Matlab Code

```
%.......................................................................%
%..................Program to find biorthogonal vectors.........................%
clc
clear all
close all

a=[3 1;1/2 3/4]; % The given vectors written as the rows of a matrix
b=inv(a) % The inverse of 'a' is stored in 'b'
d1=dot(a(1,:),b(:,1)) % dot product of first row of 'a' and first column of 'b'
d2=dot(a(1,:),b(:,2)) % dot product of first row of 'a' and second column of 'b'
d3=dot(a(2,:),b(:,1)) % dot product of second row of 'a' and first column of 'b'
d4=dot(a(2,:),b(:,2)) % dot product of second row of 'a' and second column of 'b'
```

The given vectors are stored as the rows of a matrix "a." The we found the inverse of "a" and stored in "b." We know the biorthogonal vectors are the columns of "b." The matrix "b" obtained is $\left[\frac{3}{7}\ \frac{-4}{7} \frac{-2}{7}\ \frac{12}{7}\right]$. Let us call them as $b_1 = (\frac{3}{7}, \frac{-4}{7})$ and $b_2 = (\frac{-2}{7}, \frac{12}{7})$. To verify the result, the dot product of vectors is found, and the result shows that $x_1 \perp b_2$ and $x_2 \perp b_1$.

Problem Given the discrete-time signal:

$$x[n] = \{0, 1, 2, 3, 2, 1, 0, -1, -2, -3, -2, -1\}$$

Decompose the signal using lifting scheme and reconstruct using inverse lifting scheme.

Solution

1. **Decomposition Using the Lifting Scheme:**

   (a) **Split Step:**

$$x_{\text{even}}[n] = \{0, 2, 2, 0, -2, -2\}$$

$$x_{\text{odd}}[n] = \{1, 3, 1, -1, -3, -1\}$$

   (b) **Predict Step:**

$$d[n] = x_{\text{odd}}[n] - x_{\text{even}}[n]$$

$$d[n] = \{1, 1, -1, -1, -1, 1\}$$

(c) **Update Step:**

$$s[n] = x_{\text{even}}[n] + \frac{d[n]}{2}$$

$$s[n] = \{0.5, 2.5, 1.5, -0.5, -2.5, -1.5\}$$

2. **Reconstruction Using the Inverse Lifting Scheme:**

(a) **Undo Update:**

$$x_{\text{even}}[n] = s[n] - \frac{d[n]}{2}$$

$$x_{\text{even}}[n] = \{0, 2, 2, 0, -2, -2\}$$

(b) **Undo Predict:**

$$x_{\text{odd}}[n] = d[n] + x_{\text{even}}[n]$$

$$x_{\text{odd}}[n] = \{1, 3, 1, -1, -3, -1\}$$

(c) **Merge Step:**

$$x[n] = \{0, 1, 2, 3, 2, 1, 0, -1, -2, -3, -2, -1\}$$

3. **Verification:**

$$\text{Original Signal: } x[n] = \{0, 1, 2, 3, 2, 1, 0, -1, -2, -3, -2, -1\}$$

$$\text{Reconstructed Signal: } x[n] = \{0, 1, 2, 3, 2, 1, 0, -1, -2, -3, -2, -1\}$$

Since the reconstructed signal matches the original signal, the lifting scheme and its inverse have been successfully applied. ✓

Chapter 7
Construction of Wavelets in Frequency Domain

Wavelets have a wide range of uses in signal processing. In signal analysis, they enable the detection of features at different scales that may not be visible in other domains. Wavelets can also be used for signal denoising by filtering out components with small amplitudes, and for feature extraction to identify location and amplitude of peaks or frequency content. In addition, wavelets can be used for data compression by discarding coefficients below a certain threshold. In image processing, wavelets capture both spatial and frequency information and can be utilized for denoising, feature extraction, and compression. Wavelets are also useful for time-frequency analysis of nonstationary signals and pattern recognition by comparing features with templates or training data. Finally, wavelets are applicable in speech recognition to extract features such as formants and pitch from speech signals. The construction of wavelets is necessary for these applications because it allows for the design of wavelets with specific properties that are well-suited for different applications.

7.1 Wavelet Parameters

When constructing wavelets, there are several key parameters to consider. In this section, we will discuss these parameters and their relevance in detail.

Compact Support

Compact support is an important property of wavelets that determines the size of the wavelet function. A wavelet is said to have compact support if it is nonzero only over a finite interval of time or space. This property ensures that the wavelet function

M. S. Sinith et al., *Wavelets and Multi-Resolution Analysis*,
https://doi.org/10.1007/978-3-031-92750-8_7

Table 7.1 Comparison of compact and non-compact support wavelets

Aspect	Compact support wavelets	Non-compact support wavelets
Definition	Wavelets that are nonzero over a finite interval	Wavelets that are nonzero over an infinite interval
Examples	Haar, Daubechies, Symlets, Coiflets	Meyer, Shannon, Morlet
Localization	Excellent time and frequency localization	Good frequency localization but poor time localization

decays quickly outside of its support, which is important for efficient computation and accurate analysis of signals.

The compact support property is particularly relevant in the area of signal processing and compression, where the goal is to represent signals in a compact and efficient way. Wavelets with compact support allow us to represent signals using a small number of wavelet coefficients while still preserving important features of the signal. The table 7.1 summarizes the features of wavelets with compact support and those without compact support.

For example, consider a wavelet with compact support, such as the Haar wavelet. We can efficiently represent the signal using a small number of wavelet coefficients. The Haar wavelet has a compact support of length 2, which means that it is nonzero only over the interval $[-1, 1]$. This ensures that the wavelet function decays quickly outside of its support and allows us to accurately represent the signal using only a few wavelet coefficients. On the other hand, if we use a wavelet without compact support, we may require a large number of coefficients to accurately represent the signal:

- **Support of the Scaling Function ($\varphi(t)$)**
  The scaling function $\varphi(t)$ has compact support if the scaling filter coefficients $h(n)$ have support over the interval $[N_1, N_2]$. After recursive refinement, the support of $\varphi(t)$ remains finite and is given by

$$\text{supp}(\varphi(t)) \subset [N_1, N_2] \tag{7.1}$$

  where N_1 and N_2 correspond to the boundaries of the support of $h(n)$.
- **Support of the Wavelet Function ($\psi(t)$)**
  The wavelet function $\psi(t)$ is constructed from $\varphi(t)$ using the filter coefficients $g(n)$, which also has compact support. Since $g(n)$ is nonzero only within a finite range, the support of $\psi(t)$ is given by

$$\text{supp}(\psi(t)) \subset \left[\frac{1-(N_2-N_1)}{2}, \frac{1+(N_2-N_1)}{2}\right] \tag{7.2}$$

  This shows that $\psi(t)$ has compact support with a length of $N_2 - N_1$, centered at $\frac{1}{2}$.

Vanishing Moments

Vanishing moments refer to the number of moments of the wavelet function that are equal to zero. If a wavelet function $\psi(t)$ has P vanishing moments, it means that the integral of $\psi(t)$ multiplied by t^k equals zero for all $k = 0, 1, 2, \ldots, P - 1$:

$$\int_{-\infty}^{\infty} t^k \, \psi(t) \, dt = 0 \tag{7.3}$$

This property indicates that the wavelet is orthogonal to all polynomials of degree less than $P - 1$, making it highly effective in detecting discontinuities and representing smooth signals. The number of vanishing moments P is determined by the scaling filter coefficients $h[n]$ used in the scaling equation.

A wavelet has P vanishing moments if and only if its scaling function can generate polynomials of degree up to $P - 1$. For example, the Haar wavelet has one vanishing moment ($P = 1$), making it highly discontinuous and effective in detecting sharp changes in signals. On the other hand, Daubechies wavelets with higher orders have more vanishing moments, resulting in smoother wavelet functions that can efficiently detect subtle discontinuities.

In the frequency domain, the condition for vanishing moments can be represented through the low-pass filter $H(\omega)$ associated with the scaling function:

$$\left. \frac{d^k}{d\omega^k} H(\omega) \right|_{\omega=\pi} = 0 \quad \text{for} \quad k = 0, 1, 2, \ldots, P - 1 \tag{7.4}$$

For orthogonal wavelets, if the wavelet $\psi(t)$ has P vanishing moments, its support length is at least $2P - 1$. While compact support and vanishing moments are generally independent properties, they are strongly related in the case of orthogonal wavelets. Wavelets with more vanishing moments tend to have longer support, which improves frequency localization but may reduce time localization.

Symmetry

In signal processing applications, wavelets are considered special cases of filters. A key property of certain wavelets is **symmetry** or **antisymmetry**, which leads to a desirable characteristic known as the *linear phase response*. This property ensures that all frequency components of a signal experience the same phase shift, minimizing phase distortion. This is particularly important in applications such as image and audio processing, where preserving the structure of the signal (like edges in images or transient details in audio) is crucial.

A scaling function $\varphi(t)$ is said to be **symmetric** about a point a if:

$$\varphi(a + t) = \varphi(a - t) \tag{7.5}$$

This condition implies that the function exhibits mirror symmetry around the point a. On the other hand, $\varphi(t)$ is **antisymmetric** about a point a if:

$$\varphi(a+t) = -\varphi(a-t) \tag{7.6}$$

In this case, the function has an inverted mirror symmetry, meaning the values are reflected and flipped in sign across the point a.

For **discrete filters**, which are often represented by filter coefficients $\{h_k\}$, symmetry or antisymmetry is expressed as

$$h_{a+k} = \pm h_{a-k} \tag{7.7}$$

Here:

- a represents the center of symmetry, and to achieve a linear phase response, a must be an integer.
- The $+$ sign corresponds to symmetric filters, while the $-$ sign corresponds to antisymmetric filters.

Applications Symmetric wavelets are widely used in applications such as image compression (e.g., JPEG2000), signal denoising, and edge detection due to their ability to maintain the structural integrity of the processed signals.

Regularity

Regularity refers to the number of continuous derivatives that a function possesses. Intuitively, it represents the *smoothness* of the function. In the context of wavelets, regularity indicates how smooth the wavelet function $\psi(t)$ and the scaling function $\varphi(t)$ are. A higher degree of regularity implies that the wavelet is smoother, making it better suited for analyzing signals with gradual changes and for applications where smooth approximations are required.

There is a strong connection between regularity and vanishing moments. A wavelet with more vanishing moments tends to have higher regularity. This is because vanishing moments allow the wavelet to effectively annihilate polynomials of increasing degrees, leading to smoother transitions in the wavelet function. For example, the Haar wavelet has only one vanishing moment and is discontinuous (low regularity), while Daubechies wavelets with higher orders have more vanishing moments and exhibit greater smoothness.

Regularity is an important property in applications such as image compression, denoising, and numerical analysis, where smoothness directly influences the quality of signal representation and reconstruction.

7.2 Parametric Design of Orthogonal Wavelets in the Frequency Domain

In this section, we focus on the process of determining the scaling function $\varphi(t)$ and the wavelet function $\psi(t)$ given the recursion coefficients $\{h_k\}$ and $\{g_k\}$. To facilitate this, we introduce the following frequency domain representations:

$$\begin{aligned} H(\omega) &= \frac{1}{2}\sum_k h_k e^{-ik\omega} \\ G(\omega) &= \frac{1}{2}\sum_k g_k e^{-ik\omega} \end{aligned} \tag{7.8}$$

Here, $H(\omega)$ and $G(\omega)$ are 2π-periodic functions expressed as Fourier series. When the sequences $\{h_k\}$ and $\{g_k\}$ have finite support, these functions are known as trigonometric polynomials. The conditions ensuring the convergence of these series guarantee that $H(\omega)$ and $G(\omega)$ are differentiable everywhere.

In addition to the Fourier series representation, we define the following Z transform representations:

$$\begin{aligned} H(z) &= \frac{1}{2}\sum_k h_k z^{-k} \\ G(z) &= \frac{1}{2}\sum_k g_k z^{-k} \end{aligned} \tag{7.9}$$

These functions represent the actual Z transform of the filters h and g, where z is a complex variable. The negative exponent z^{-k} reflects the time-shifting property commonly used in digital signal processing. The Z transform is particularly useful for analyzing the stability and frequency response of discrete-time systems.

A key relationship connects the frequency domain and the Z-domain representations:

$$\begin{aligned} H(\omega) &= H(z)\big|_{z=e^{i\omega}} \\ G(\omega) &= G(z)\big|_{z=e^{i\omega}} \end{aligned} \tag{7.10}$$

This shows that $H(\omega)$ and $G(\omega)$ can be obtained by evaluating the Z transforms $H(z)$ and $G(z)$ on the unit circle in the complex plane ($|z| = 1$).

The purpose of introducing these functions is to provide different perspectives on the same set of information. The coefficients $\{h_k\}$ and $\{g_k\}$, as well as the functions $H(\omega)$, $G(\omega)$, and their Z transforms $H(z)$ and $G(z)$, are mathematically equivalent—they represent the same data in different forms. Depending on the context, one representation may offer computational or analytical advantages over the others. This flexibility is similar to switching between time and frequency domains using the Fourier transform in signal processing.

To further analyze the scaling function $\varphi(t)$, we take the Fourier transform of the recursion relation (which is another name for refinement relations):

$$\varphi(t) = \sqrt{2} \sum_k h_k \varphi(2t - k) \tag{7.11}$$

Applying the Fourier transform to both sides, we obtain

$$\hat{\varphi}(\omega) = \sqrt{2} \sum_k h_k \cdot \frac{1}{2} e^{-ik\omega/2} \hat{\varphi}\left(\frac{\omega}{2}\right) \tag{7.12}$$

This simplifies to

$$\hat{\varphi}(\omega) = \left(\frac{1}{\sqrt{2}} \sum_k h_k e^{-ik\omega/2} \right) \hat{\varphi}\left(\frac{\omega}{2}\right) \tag{7.13}$$

Recognizing the term in parentheses as $H\left(\frac{\omega}{2}\right)$, the Fourier transform of the scaling filter evaluated at $\omega/2$, we get

$$\hat{\varphi}(\omega) = H\left(\frac{\omega}{2}\right) \hat{\varphi}\left(\frac{\omega}{2}\right) \tag{7.14}$$

This recursive relationship is fundamental to the multiresolution analysis framework.

Repeating this process iteratively, we obtain

$$\hat{\varphi}(\omega) = H\left(\frac{\omega}{2}\right) H\left(\frac{\omega}{4}\right) \hat{\varphi}\left(\frac{\omega}{4}\right) \tag{7.15}$$

Continuing this process indefinitely leads to

$$\hat{\varphi}(\omega) = \prod_{j=1}^{\infty} H\left(\frac{\omega}{2^j}\right) \tag{7.16}$$

This expression represents the scaling function in the frequency domain as an infinite product of the frequency response of the scaling filter evaluated at progressively smaller frequencies.

Repeating the recursive relation for the Fourier transform of the scaling function, we obtain

$$\hat{\varphi}(\omega) = H\left(\frac{\omega}{2}\right) H\left(\frac{\omega}{4}\right) \hat{\varphi}\left(\frac{\omega}{4}\right) = \cdots = \left(\prod_{j=1}^{\infty} H\left(\frac{\omega}{2^j}\right) \right) \tag{7.17}$$

The conditions ensuring the convergence of this infinite product are typically referred to as admissibility conditions. Since $H(\omega)$ is differentiable at $\omega = 0$ and $H(0) = 1$, we have the following inequality for ω near 0:

$$|H(\omega)| \leq 1 + c|\omega| \leq e^{c|\omega|} \tag{7.18}$$

where c is a positive constant.

Conditions for Orthonormality in the Frequency Domain

To ensure that the scaling function $\varphi(t)$ and the wavelet function $\psi(t)$ form an orthonormal basis, certain conditions must be satisfied in the frequency domain. These conditions involve the frequency responses of the scaling filter $H(\omega)$ and the wavelet filter $G(\omega)$.

The orthogonality conditions for the scaling and wavelet functions can be expressed as follows:

$$H(0) = 1 \tag{12a}$$

$$|H(\omega)|^2 + |H(\omega + \pi)|^2 = 1 \tag{12b}$$

$$H(\pi) = 0 \tag{12c}$$

$$G(0) = 0 \tag{12d}$$

$$|G(\omega)|^2 + |G(\omega + \pi)|^2 = 1 \tag{12e}$$

$$H(\omega)\overline{G(\omega)} + H(\omega + \pi)\overline{G(\omega + \pi)} = 0 \tag{12f}$$

These conditions ensure the following:

- Low-Frequency Normalization: $H(0) = 1$ ensures the scaling function has a nonzero integral.
- Energy Conservation for the Scaling Filter: The total energy is preserved across frequency bands.
- High-Frequency Nulling: $H(\pi) = 0$ ensures no overlap with the wavelet function at high frequencies.
- Wavelet Vanishing at Zero Frequency: $G(0) = 0$ ensures the wavelet has zero mean.

Polyphase Filtering

Polyphase filtering provides an efficient framework for implementing filter banks, which are fundamental in wavelet-based signal processing. This approach decomposes the filters into subcomponents that handle even and odd samples separately, enhancing computational efficiency and offering insights into the structural properties of wavelet transforms.

When a discrete-time signal $s = \{s_k\}_{k\in\mathbb{Z}}$ passes through both low-pass and high-pass filters, it undergoes dyadic downsampling (denoted by $\downarrow 2$), where odd-indexed samples are discarded, and the remaining even-indexed samples are relabeled. This process produces a level-1 approximation signal

$$a = \{a_k\}_{k\in\mathbb{Z}} \tag{7.19}$$

and a level-1 detail signal

$$b = \{b_k\}_{k\in\mathbb{Z}}, \tag{7.20}$$

defined as

$$a_k = \sum_{\ell\in\mathbb{Z}} h_\ell s_{2k-\ell}, \tag{7.21}$$

$$b_k = \sum_{\ell\in\mathbb{Z}} g_\ell s_{2k-\ell} \tag{7.22}$$

This procedure can be iteratively applied to the approximation signal a to generate approximation and detail signals at higher levels. Repeating this process allows the decomposition of the original signal into coarser approximation and detail signals up to a maximum level $j_{\max}$, which is constrained by the finite length of the original signal.

An alternative and efficient approach to implementing the above process is through polyphase filtering. Here, the signal s is split into its even and odd samples, denoted by

$$s_{\text{even}} = \{s_{2k}\}_{k\in\mathbb{Z}} \tag{7.23}$$

and

$$s_{\text{odd}} = \{s_{2k+1}\}_{k\in\mathbb{Z}} \tag{7.24}$$

respectively. These two phases are processed jointly using a 2×2 polyphase matrix $P(z)$, defined as

$$P(z) = \begin{pmatrix} H_{\text{even}}(z) & H_{\text{odd}}(z) \\ G_{\text{even}}(z) & G_{\text{odd}}(z) \end{pmatrix} \tag{7.25}$$

where:

$$H_{\text{even}}(z) = \sum_{k \in \mathbb{Z}} h_{2k} z^{-k}, \tag{7.26}$$

$$H_{\text{odd}}(z) = \sum_{k \in \mathbb{Z}} h_{2k+1} z^{-k}, \tag{7.27}$$

$$G_{\text{even}}(z) = \sum_{k \in \mathbb{Z}} g_{2k} z^{-k}, \tag{7.28}$$

$$G_{\text{odd}}(z) = \sum_{k \in \mathbb{Z}} g_{2k+1} z^{-k} \tag{7.29}$$

The output signals a and b can be obtained by applying this polyphase matrix to the even and delayed odd samples of the signal s. This formulation offers computational advantages, particularly in real-time applications, and provides a clear structure for analyzing filter properties, such as orthogonality and perfect reconstruction.

A polyphase matrix $P(z)$ facilitates perfect reconstruction of the original signal if and only if $P(z)$ is invertible, meaning $P(z)^{-1}$ exists. This condition is met when the filters $H(z)$ and $G(z)$ are power complementary, satisfying:

$$H(z)H(z^{-1}) + G(z)G(z^{-1}) = 2 \tag{7.30}$$

Power complementary filters ensure that the filter bank decomposition is lossless, preserving all the information from the original signal and allowing exact reconstruction after decomposition. This property is closely related to the orthogonality of the wavelet basis functions generated by the filters.

For orthogonality to hold, it is critical that the polyphase matrix $P(z)$ behaves as an all-pass filter, meaning that it is unitary on the unit circle $|z| = 1$:

$$P(z)P^*(z) = I \tag{7.31}$$

where $P^*(z)$ is the Hermitian transpose (complex conjugate transpose) of $P(z)$, and I is the identity matrix. This condition ensures that no information is lost during filtering and that the decomposition is energy preserving.

Designing orthogonal wavelets often involves optimizing certain criteria over a parameterized family of lossless polyphase filters, avoiding the need to impose complex nonlinear algebraic constraints explicitly. This approach enhances numerical stability and improves the performance of optimization algorithms used in filter design.

FIR Filter Characteristics and Parameterization of Wavelets

Having established the foundation of polyphase filtering and its role in wavelet decomposition, we now delve into the characteristics of finite impulse response (FIR) filters, which form the core components of these filter banks. FIR filters are preferred in wavelet-based signal processing due to their inherent stability, linear phase properties, and efficient implementation. These filters are characterized by having a finite number of nonzero coefficients, ensuring compact support, which is crucial for localized signal analysis in both time and frequency domains.

A linear time-invariant (LTI) filter $F(z)$ is said to be causal if it depends only on the present and past input values. This is mathematically represented as

$$F(z) = \sum_{k \in \mathbb{Z}} f_k z^{-k}, \quad \text{with } f_k = 0 \text{ for all } k < 0 \tag{7.32}$$

For wavelet design, compact support is a desirable property, meaning the filter has a finite impulse response. This ensures the wavelet is well localized, capturing local features efficiently. In this context, the low-pass $H(z)$ and high-pass $G(z)$ filters are designed as FIR filters, taking the form of moving average (MA) systems rather than more general ARMA models. These filters have all their poles at the origin, represented as

$$H(z) = h_0 + h_1 z^{-1} + \cdots + h_N z^{-N}, \quad G(z) = g_0 + g_1 z^{-1} + \cdots + g_N z^{-N} \tag{7.33}$$

Here, N is the filter order. For orthogonal FIR filter banks, orthogonality conditions imply a specific relationship between the low-pass and high-pass filters, known as the alternating flip construction. This relationship ensures perfect reconstruction in filter banks, where the high-pass filter coefficients are derived from the low-pass coefficients using:

$$g_k = (-1)^k h_{N-k}, \quad (k = 0, 1, \ldots, N) \tag{7.34}$$

The orthonormality condition for the scaling function basis $\{\varphi(t-m)\}$ requires the filter order N to be odd, typically $N = 2n - 1$. This constraint ensures proper wavelet construction, where the scaling function $\varphi(t)$ and the wavelet $\psi(t)$ are both supported within a finite interval, enhancing time-frequency localization.

Parameterization of Wavelets as Lossless FIR Polyphase Filters

To optimize wavelet design, parameterization of wavelets as lossless FIR polyphase filters is employed. This approach allows efficient manipulation of filter properties, enabling the design of orthogonal wavelets with desired characteristics. The polyphase matrix $P(z)$, associated with the filter bank, can be factored into a product of rotation matrices and elementary delay matrices, reflecting its lossless nature:

$$P(z) = R(\theta_n)\Lambda(z)R(\theta_{n-1})\Lambda(z)\cdots\Lambda(z)R(\theta_1)\Lambda(-1) \tag{7.35}$$

where:

$$R(\theta) = \begin{pmatrix} \cos(\theta) & -\sin(\theta) \\ \sin(\theta) & \cos(\theta) \end{pmatrix} \tag{7.36}$$

$$\Lambda(z) = \begin{pmatrix} 1 & 0 \\ 0 & z^{-1} \end{pmatrix} \tag{7.37}$$

This parameterization introduces angular parameters $\theta_1, \theta_2, \ldots, \theta_n$, which control the filter properties. The matrix $\Lambda(-1)$ ensures consistency with the sign convention required by the alternating flip construction.

An important advantage of this factorization is that it enables the design of wavelets through lattice filter implementations, improving computational efficiency and numerical stability. The filter coefficients h_k and g_k can be recursively computed from the angular parameters, simplifying optimization routines for wavelet design.

7.3 Design of Matched Wavelets

Matched wavelets are designed to closely match the features of the signals of interest. As a result, they can provide better representation and preservation of important signal details while suppressing noise and irrelevant components. This leads to an improved signal-to-noise ratio and enhanced signal quality. In compression applications, matched wavelets can often achieve higher compression ratios compared to generic wavelets. This is because they concentrate their representation power on the essential information of the signal, allowing better compression with fewer coefficients. Matched wavelets can be tailored to enhance the detection of specific features in the signal. For example, in image processing, matched wavelets can highlight edges or textures, making them particularly useful in tasks like edge detection or texture analysis. When analyzing signals using wavelet transforms, using matched wavelets can lead to more accurate results. By capturing the essential characteristics of the signal, matched wavelets enable more precise feature extraction and analysis.

In some cases, matched wavelets can be designed to have fewer vanishing moments than standard wavelets, reducing the computational complexity of the wavelet transform while maintaining good performance. Matched wavelets can be tailored to specific applications, data types, or signal classes. This customization allows engineers and researchers to design wavelets that are optimized for their particular use case, leading to better performance and more efficient processing.

Matched Wavelet by Spectral Matching

In this section, we discuss the algorithm proposed by Chapa et al. (2000) for designing wavelets that are matched to a specified signal. This technique assumes that the wavelets are bandlimited. Two sets of equations are developed to design wavelets directly from the signal of interest:

1. Derive the expression for the continuous matched wavelet spectrum amplitudes.
2. Develop a direct discrete algorithm for calculating close approximations to the optimal complex wavelet spectrum.

The matching process was performed on the magnitude and phase independently. When designing matched wavelets directly from a given signal, the wavelet spectrum $\Psi(\omega)$ is derived first. However, there is no closed-form expression to obtain the scaling function spectrum $\Phi(\omega)$ directly from $\Psi(\omega)$. The algorithm focuses on determining both the spectral amplitude and phase for $\Phi(\omega)$ and $\Psi(\omega)$.

Amplitude Determination
For a given signal $f(x)$, the amplitude of the matched wavelet function, $|\Psi(\omega)|$, is computed. The amplitude of the scaling function, $|\Phi(\omega)|$, is then obtained using the relationship:

$$|\Phi(\omega)|^2 = \sum_{k=1}^{\infty} |\Psi(\omega + 2\pi k)|^2 \tag{7.38}$$

This ensures that the scaling function captures the complementary spectral content not covered by the wavelet function.

Phase Determination
The phases of both $\Psi(\omega)$ and $\Phi(\omega)$ are obtained through an algorithm. Combining the amplitude and phase information provides the complete spectra for $\Psi(\omega)$ and $\Phi(\omega)$, which are used to derive the corresponding filter coefficients $h(n)$ (low pass) and $g(n)$ (high pass).

Orthonormality Conditions
To guarantee orthonormality of the wavelet function $\Psi(\omega)$, the following condition must be satisfied:

$$\sum_{m=-\infty}^{\infty} \sum_{k=1}^{\infty} \left(|\Phi(\omega + 2\pi m)|^2 + |\Psi(\omega + 2\pi k)|^2 \right) = 1 \tag{7.39}$$

This condition ensures that the energy of the scaling and wavelet functions is preserved, satisfying the orthonormality required in wavelet analysis.

Multiresolution Analysis (MRA) Conditions
To meet MRA conditions, the matched wavelets must satisfy specific criteria proven in related theorems. These include ensuring that:

$$\Phi(\omega) \neq 0 \quad \text{for} \quad |\omega| < \frac{4\pi}{3} \tag{7.40}$$

and

$$\Psi(\omega) = 0 \quad \text{for} \quad |\omega| < \pi \tag{7.41}$$

This guarantees the proper scaling and wavelet functions necessary for effective signal decomposition in MRA.

7.4 Matching Spectrum Amplitude

The spectral sampling resolution of the scaling function $\Phi(\omega)$ and the wavelet function $\Psi(\omega)$ is defined based on the number of spectral samples. The resolutions are given as

$$\Delta\omega_{\Phi} = \frac{2\pi}{2l + 1}, \quad \Delta\omega_{\Psi} = \frac{2\pi}{2M} \tag{7.42}$$

where l and M define the number of frequency components used in the scaling and wavelet functions, respectively. Given these resolutions, the spectra are discretized, leading to sampled versions of the scaling and wavelet functions, denoted as $\Phi(n\Delta\omega_{\Phi})$ and $\Psi(n\Delta\omega_{\Psi})$. This discretization allows the derivation of computationally efficient filter coefficients while preserving the spectral characteristics of the wavelet transform.

To ensure that the wavelet maintains energy conservation and orthonormality conditions in the discrete domain, the previously established continuous equations are rewritten in discretized form. The total energy constraint for the scaling and wavelet functions is expressed as

$$\sum_{p=0}^{l} \left(\Phi^2(p\Delta\omega_\Phi) + \Psi^2(p\Delta\omega_\Psi) \right) = \pi \tag{7.43}$$

Additionally, the orthonormality condition ensuring completeness of the wavelet representation in the decomposition process is given by

$$\sum_{m=-\infty}^{\infty} \sum_{p=0}^{l+M} \Psi^2(n + m\Delta\omega_\Psi) = 1 \tag{7.44}$$

These constraints ensure that the discrete representation retains the fundamental properties required for wavelet-based signal decomposition.

The multiresolution analysis (MRA) framework imposes further constraints on the frequency bands allowed for the wavelet function. Specifically, the range of permissible spectral components is defined as

$$\frac{2M}{3} < p + n + m < 2M + \frac{2}{3} \tag{7.45}$$

This ensures that the wavelet function is localized within the desired frequency range, aligning with the decomposition requirements of the wavelet transform.

Since the wavelet function $\Psi(\omega)$ is assumed to be real-valued, its spectrum must be an even function. This property simplifies computations, as only positive frequency indices need to be considered. The sampled power spectrum of the wavelet is represented as

$$Y(k) = |\Psi(k\Delta\omega)|^2 \tag{7.46}$$

To determine $Y(k)$, a system of linear equations is formulated:

$$\sum_{i=1}^{k} \alpha_{ij} = Y(k) \tag{7.47}$$

which can be compactly expressed in vector form as

$$AY = 1 \tag{7.48}$$

where A is a matrix with elements $\alpha_{ij} \in \{0, 1, 2\}$, and 1 is a unity vector of length L. This formulation enables the estimation of the wavelet spectrum using linear algebraic techniques.

To minimize the normalized mean square error (NMSE) between the sampled power spectrum of the matched wavelet $Y(k)$ and the reference power spectrum $W(k)$, the following error function is defined:

$$E = \frac{\sum_k (W(k) - Y(k))^2}{\sum_k W(k)} \tag{7.49}$$

Applying Lagrangian multipliers and introducing a weighting parameter a, the minimization process results in the optimal solution for Y as

$$Y = \left(A^T A + \frac{1}{a} I\right)^{-1} A^T W \tag{7.50}$$

Here:

- A is the coefficient matrix derived from the linear system.
- A^T is the transpose of A.
- I is the identity matrix.
- a is the weighting parameter controlling the regularization.

Further simplification of the solution, considering the Lagrangian formulation, yields

$$Y = \frac{1}{a} \left(A^T A + \frac{1}{a} I\right)^{-1} A^T W - \frac{1}{a} W \tag{7.51}$$

This equation ensures that the error E is minimized effectively, balancing the influence of both the target spectrum W and the structural constraints imposed by matrix A.

The optimized power spectrum of the matched wavelet is then computed as

$$|\Psi(k\Delta\omega)|^2 = Y(k) \tag{7.52}$$

Finally, using the discretized form of the energy condition, the amplitude of the scaling function $\Phi(\omega)$ can be derived, ensuring that the designed wavelet satisfies the required orthonormality and multiresolution analysis (MRA) conditions.

Matching Spectrum Phase

Direct matching of the phase is not feasible. Instead, we address the problem through group delay. We start by expressing the group delay of $\Psi(\omega)$ in terms of the group delay of the scaling function spectrum, $\Phi(\omega)$. We have

$$\Psi(2\omega) = e^{-i\omega} \frac{\overline{\Phi(2\omega + 2\pi)}}{\overline{\Phi(\omega + \pi)}} \Phi(\omega), \tag{7.53}$$

which leads to the phase $\theta_\Psi(\omega)$ being:

$$\theta_\Psi(\omega) = -\frac{\omega}{2} - \theta_\Phi(\omega + 2\pi) + \theta_\Phi\left(\frac{\omega}{2} + \pi\right) + \theta_\Phi\left(\frac{\omega}{2}\right), \tag{7.54}$$

where $\theta_\Psi(\omega)$ and $\theta_\Phi(\omega)$ denote the phases of Ψ and Φ, respectively. Let Λ_Ψ and Λ_Φ represent the negative of the group delays, and define $\Gamma_\Psi(\omega) = \Lambda_\Psi(\omega) + \frac{1}{2}$. We then derive

$$\Psi(\omega) = \prod_{m=1}^{\infty} H\left(\frac{\omega}{2^m}\right) \tag{7.55}$$

and

$$\Psi(\omega) = e^{-i\frac{\omega}{2}}\overline{H\left(\frac{\omega}{2} + \pi\right)} \prod_{m=2}^{\infty} H\left(\frac{\omega}{2^m}\right), \tag{7.56}$$

where $H(\omega)$ is 2π-periodic. Taking the phase of both sides gives

$$\theta_\Phi = \sum_{m=1}^{\infty} \theta_H\left(\frac{\omega}{2^m}\right), \tag{7.57}$$

and

$$\theta_\Psi = -\frac{\omega}{2} - \theta_H\left(\frac{\omega}{2} + \pi\right) + \sum_{m=2}^{\infty} \theta_H\left(\frac{\omega}{2^m}\right). \tag{7.58}$$

Taking the derivative yields the negative group delays:

$$\begin{aligned} \Lambda_\Phi(\omega) &= \sum_{m=1}^{\infty} 2^{-m}\lambda\left(\frac{\omega}{2^m}\right), \\ \Gamma_\Psi(\omega) &= \Lambda_\Psi(\omega) + \frac{1}{2} = -\frac{1}{2}\lambda\left(\frac{\omega}{2} + \pi\right) + \sum_{m=2}^{\infty} 2^{-m}\lambda\left(\frac{\omega}{2^m}\right), \end{aligned} \tag{7.59}$$

where $\Lambda_\Phi(\omega) = \frac{d\theta_\Phi(\omega)}{d\omega}$, $\Lambda_\Psi(\omega) = \frac{d\theta_\Psi(\omega)}{d\omega}$, and $\lambda(\omega) = \frac{d\theta_H(\omega)}{d\omega}$ are 2π-periodic and even functions. To solve the issue of matching the group delays directly, model $\lambda(\omega)$ as a polynomial of even order R over one period:

$$\lambda_T(\omega) = \sum_{r=0}^{R/2} c_r\omega^{2r} \prod\left(\frac{\omega}{2\pi}\right) \tag{7.60}$$

where

$$\prod(\omega) = \begin{cases} 1 & \text{if } -\frac{1}{2} \leq \omega < \frac{1}{2} \\ 0 & \text{otherwise} \end{cases} \tag{7.61}$$

Replicating $\lambda_T(\omega)$ every 2π, we get

$$\lambda(\omega) = \sum_{k=-\infty}^{\infty} \lambda_T(\omega - 2\pi k) = \sum_{k=-\infty}^{\infty} \sum_{r=0}^{R/2} c_r(\omega - 2\pi k)^{2r} \prod\left(\frac{\omega - 2\pi k}{2\pi}\right) \tag{7.62}$$

Let N be the number of samples in $F(n\Delta\omega)$ with sampling frequency $\Delta\omega = 2\pi/T$. The discrete form is given by

$$\lambda(n) = \sum_{r=0}^{R/2} c_r \sum_{k=-P/2}^{P/2-1} (n - kT)^{2r} \prod\left(\frac{n - kT}{T}\right) \tag{7.63}$$

This yields matrix equations for Γ_Ψ and Λ_Φ:

$$\begin{aligned} \Gamma_\Psi &= D_\Psi c, \\ \Lambda_\Phi &= D_\Phi c \end{aligned} \tag{7.64}$$

where:

$$\begin{aligned} D_\Psi &= -\frac{1}{2} B_{(q+T)/2} + \sum_{m=2}^{\infty} 2^{-m} B_{(q/2^m)}, \\ D_\Phi &= \sum_{m=1}^{\infty} 2^{-m} B_{(q/2^m)} \end{aligned} \tag{7.65}$$

To find the closest match to the desired signal's group delay, Γ_F, in a least squares sense, let γ be the unweighted error:

$$\gamma = \sum_{n=-N/2}^{N/2-1} (\Gamma_F(n) - \Gamma_\Psi(n))^2 \tag{7.66}$$

Since phase matching needs only to be accurate in the passband, we weight the error function by a normalized weighting function $\Omega(n) = Y(n)/\sum Y(n)$, where $Y(n)$ are elements of $\mathbf{Y}$:

$$\gamma_\Omega = \sum_{n=-N/2}^{N/2-1} [\Omega(n) (\Gamma_F(n) - \Gamma_\Psi(n))]^2 . \tag{7.67}$$

Rewriting in vector notation:

$$\gamma = \left(\overline{\Gamma_F} - \overline{D_\Psi c}\right)^T \left(\overline{\Gamma_F} - \overline{D_\Psi c}\right), \tag{7.68}$$

where $\overline{\Gamma_F}$ and $\overline{D_\Psi}$ are obtained by considering the nonzero entries of weighted Γ_F and D_Ψ. Setting $\nabla_c \gamma = 0$, the vector $\hat{c}$ that minimizes γ is

$$\hat{c} = \left(\overline{D_\Psi}^T \overline{D_\Psi}^{-1} \overline{D_\Psi}^T \overline{\Gamma_F}\right) \tag{7.69}$$

With this best estimate of c, the group delay of the wavelet can be obtained as

$$\Gamma_\Psi = D_\Psi \hat{c} \tag{7.70}$$

The best estimates of λ, Λ_Ψ, and Λ_Φ are

$$\begin{aligned} \lambda &= B\hat{c}, \\ \Lambda_\Psi &= \left(D_\Psi \hat{c} - \overline{D_\Psi \hat{c}}\right) - \frac{\Delta\omega}{2}, \\ \Lambda_\Phi &= D_\Phi \hat{c} - \overline{D_\Phi \hat{c}} \end{aligned} \tag{7.71}$$

where $\overline{D_\Psi \hat{c}}$ and $\overline{D_\Phi \hat{c}}$ are the means of $D_\Psi \hat{c}$ and $D_\Phi \hat{c}$, respectively. Both Λ_Φ and Λ_Ψ can then be summed to obtain the discrete phases of Φ and Ψ, which, when combined with the magnitudes, provide the complete wavelet representation.

7.5 Solved Problems

Problem Consider the parametric approach to designing wavelets by matching their spectra to a desired spectrum. Let **W** and **Y** be vectors containing the samples of $|F(k\Delta\omega)|^2$ and $|\Phi(k\Delta\omega)|^2$, respectively, where $F(\omega)$ is the spectrum of the desired signal and $\Psi(\omega)$ is the matched wavelet spectrum:

1. Derive the optimal parameter a given the error function:

$$E = \frac{(W - aY)^T (W - aY)}{W^T W},$$

and the optimal wavelet power spectrum:

$$Y = \frac{1}{a} W + A^T (AA^T)^{-1} \left(1 - \frac{1}{a} AW\right)$$

where

$$a = \frac{1^T(AA^T)^{-1}AW}{1^T(AA^T)^{-1}1}$$

Show the steps clearly to derive the expression for the optimal parameter a.

We are given the error function for matching the spectrum amplitude as

$$E = \frac{(W - aY)^T(W - aY)}{W^T W}$$

where:

- W is the vector of samples of $|F(k\Delta\omega)|^2$ (the desired signal spectrum).
- Y is the vector of samples of $|\Phi(k\Delta\omega)|^2$ (the spectrum of the matched wavelet).

Solution We expand the quadratic form in the numerator:

$$(W - aY)^T(W - aY) = W^T W - 2aW^T Y + a^2 Y^T Y$$

Substituting this into the error function:

$$E(a) = \frac{W^T W - 2aW^T Y + a^2 Y^T Y}{W^T W}$$

Now, we rewrite the error function as

$$E(a) = 1 - 2a\frac{W^T Y}{W^T W} + a^2\frac{Y^T Y}{W^T W}$$

To minimize $E(a)$, we need to find the value of a that minimizes the error. This involves taking the derivative of $E(a)$ with respect to a and setting it equal to zero. The derivative of $E(a)$ with respect to a is

$$\frac{dE}{da} = -2\frac{W^T Y}{W^T W} + 2a\frac{Y^T Y}{W^T W}$$

Setting this derivative equal to zero to minimize $E(a)$:

$$0 = -2\frac{W^T Y}{W^T W} + 2a\frac{Y^T Y}{W^T W}$$

Solving for a:

$$a = \frac{W^T Y}{Y^T Y}$$

This gives the optimal parameter a. The optimal wavelet power spectrum Y is given by

$$Y = \frac{1}{a}W + A^T(AA^T)^{-1}\left(1 - \frac{1}{a}AW\right)$$

Substituting the expression for a into this equation:

$$Y = \frac{Y^TY}{W^TY}W + A^T(AA^T)^{-1}\left(1 - \frac{Y^TY}{W^TY}AW\right)$$

This simplifies to the final form of the optimal wavelet power spectrum.

Problem Consider the weighted error function for phase matching:

$$\gamma_\Omega = \sum_{n=-N/2}^{N/2-1} [\Omega(n)\left(\Gamma_F(n) - \Gamma_\Psi(n)\right)]^2,$$

where

$$\Omega(n) = \frac{Y(n)}{\sum Y(n)}.$$

1. Derive the expression for the vector $\hat{c}$ that minimizes the weighted error γ_Ω. Show the steps to solve for $\hat{c}$ using the least squares method.
2. Implement a numerical example to estimate $\hat{c}$ using given sample values of $\Gamma_F(n)$, $\Omega(n)$, and $\Gamma_\Psi(n)$. Provide a brief explanation of your calculations.

Solution The weighted error function can be written as

$$\gamma_\Omega(\hat{c}) = \sum_{n=-N/2}^{N/2-1} \Omega(n)\left[\Gamma_F(n) - \Gamma_\Psi(n, \hat{c})\right]^2,$$

where $\Gamma_\Psi(n, \hat{c})$ is a function of $\hat{c}$, and we aim to minimize this error with respect to $\hat{c}$. Let the residual at each point be

$$r(n, \hat{c}) = \Gamma_F(n) - \Gamma_\Psi(n, \hat{c}),$$

so the error function becomes

$$\gamma_\Omega(\hat{c}) = \sum_{n=-N/2}^{N/2-1} \Omega(n)\left[r(n, \hat{c})\right]^2.$$

To minimize $\gamma_\Omega(\hat{c})$, we take the derivative of the error function with respect to $\hat{c}$, set it to zero, and solve for $\hat{c}$. First, differentiate $\gamma_\Omega(\hat{c})$ with respect to $\hat{c}$:

$$\frac{\partial \gamma_\Omega(\hat{c})}{\partial \hat{c}} = \sum_{n=-N/2}^{N/2-1} \Omega(n) \cdot 2\left[\Gamma_F(n) - \Gamma_\Psi(n, \hat{c})\right] \cdot \frac{\partial \Gamma_\Psi(n, \hat{c})}{\partial \hat{c}}$$

To minimize, we set the derivative equal to zero:

$$\sum_{n=-N/2}^{N/2-1} \Omega(n) \cdot \left[\Gamma_F(n) - \Gamma_\Psi(n, \hat{c})\right] \cdot \frac{\partial \Gamma_\Psi(n, \hat{c})}{\partial \hat{c}} = 0$$

This equation represents the necessary condition for minimizing $\gamma_\Omega(\hat{c})$. The exact solution for $\hat{c}$ depends on the specific form of $\Gamma_\Psi(n, \hat{c})$, which could be linear or nonlinear in terms of $\hat{c}$.

Chapter 8
Applications of Wavelet Transform

This chapter provides a comprehensive exploration of wavelet transforms, emphasizing their applications in denoising and compression. Wavelet denoising is a signal processing technique that uses wavelet transformations to reduce noise in a signal while preserving important features. Wavelet-based compression is a technique used to reduce the storage space required for representing and transmitting signals or images while preserving essential information. Unlike traditional compression methods, wavelet compression exploits the mathematical properties of wavelet transforms to achieve a more efficient representation of signals.

8.1 Wavelet Denoising

Denoising removes noise present and retains the signal present regardless of the frequency content of the signal. The wavelet transforms of real-world signals tend to produce low value coefficients at finer scale, and the finer scales capture higher frequency details or rapid changes in the signal. In many natural signals, these high-frequency components often have lower magnitudes compared to the overall signal. As a result, during the wavelet transform, coefficients at finer scales tend to represent subtle variations or noise in the signal, which may have lower values.

The wavelet transform helps in separating the signal into different scales, making it easier to identify and manipulate features at various levels of detail. The lower magnitude coefficients at finer scales may correspond to noise or fine-scale details that can be thresholded or removed during denoising processes to enhance the overall signal quality.

Wavelet denoising typically involves the following steps (Fig. 8.1):

Decomposition Decompose the signal into different scales using a wavelet transform. This results in approximation coefficients (representing the coarse structure of the signal) and detail coefficients (representing fine-scale details, including noise).

M. S. Sinith et al., *Wavelets and Multi-Resolution Analysis*,
https://doi.org/10.1007/978-3-031-92750-8_8

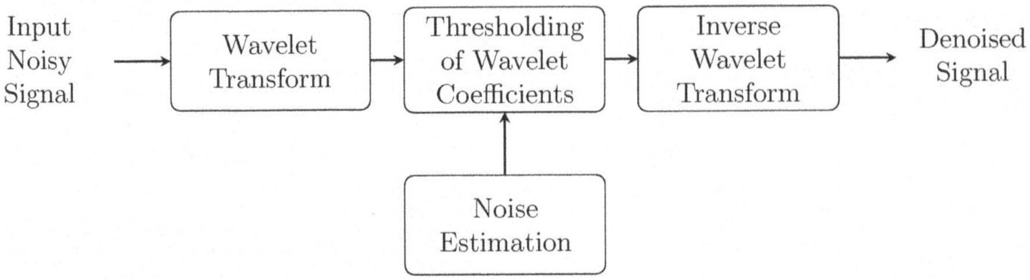

Fig. 8.1 Block diagram for denoising signal using wavelet transform

Thresholding Threshold the detail coefficients to remove or reduce the influence of noise. Thresholding involves setting small coefficients to zero or reducing their magnitude based on a certain threshold value. Common thresholding techniques include hard thresholding and soft thresholding.

Reconstruction Reconstruct the denoised signal using the modified coefficients. This involves combining the thresholded approximation and detail coefficients. The reconstructed signal should ideally retain important features while suppressing noise.

Noise Estimation

In this subsection, we explore the estimation of noise within a signal. The noisy output, represented as $X[n]$, can be expressed in terms of the input signal $f[n]$ and the noise component $W[n]$. The noise $W[n]$ is modeled as a random vector following a Gaussian distribution.

$$X[n] = f[n] + W[n] \tag{8.1}$$

The estimation of $f[n]$ from the noisy signal $X[n]$ is represented as $\tilde{F} = DX[n]$, where D is the decision operator designed to minimize the error $f - \tilde{F}$. For one dimensional signals, the mean square error is commonly used to measure this error, and the risk of the estimation is given by

$$r(D, f) = E\left\{\|f - \tilde{F}\|^2\right\} \tag{8.2}$$

The decision operator D is optimized using prior information about the signal. Two prevalent estimation methods are Bayes estimation and minimax estimation.

Bayes Estimation
In Bayes estimation, the unknown signal f, denoted by a random vector F, is assumed to follow a probability distribution α, known as the prior distribution. The

noisy output can be rewritten as

$$X[n] = F[n] + W[n] \quad \text{for} \quad 0 \leq n < N \tag{8.3}$$

Here, the noise W is assumed to be independent of F for all n. The distribution of the measurement X is the joint distribution of F and W, known as the posterior distribution. Thus, $\tilde{F} = DX$ is an estimator of F from measurement X, and the Bayes risk with respect to the prior distribution α of the signal is

$$r(D, \alpha) = E_\alpha\{r(D, F)\} = E\left\{\|F - \tilde{F}\|^2\right\} = \sum_0^{N-1} E\left\{|F[n] - \tilde{F}[n]|^2\right\} \tag{8.4}$$

An estimator $\tilde{F}$ is considered a Bayes estimator if it minimizes the Bayes risk among all estimators. The risk function is typically determined by measuring the distance between the estimator $\tilde{F}$ and the unknown signal F. The mean square error is commonly adopted due to its simplicity.

Minimax Estimation

When prior information about a signal is available, but its probability distribution is unknown, minimax estimation is employed. This method finds a "good" estimator whose maximal risk is minimal among all estimators. The prior information forms a signal set Θ, but this set does not specify the probability distribution of signals in Θ.

For a signal $f \in \Theta$, the noisy output is given by $X[n] = f[n] + W[n]$. The risk of estimation $\tilde{F} = DX$ is $r(D, f) = E\left\{\|DX - f\|^2\right\}$. Since the precise risk cannot be calculated due to the unknown distribution, the maximum risk is

$$r(D, \Theta) = \sup_{f \in \Theta} E\left\{\|DX - f\|^2\right\} \tag{8.5}$$

In minimax estimation, the minimax risk is the lower bound of the risk $r(D, \Theta)$ across all possible operators D:

$$r_n(D, \Theta) = \inf_{D \in \mathrm{O}_n} r(D, \Theta) \tag{8.6}$$

Here, O_n denotes the set of all operators.

Threshold Estimation

The estimated noise level is known as the threshold. Signals are defined as values greater than the threshold, while noises are identified as smaller values. The noise estimator can also be applied when the noisy output is decomposed into a selected basis, as white noise remains white noise across all bases. There are primarily two types of thresholding functions: soft thresholding and hard thresholding.

In the orthogonal basis $B = \{g_m\}_{0 \le m < N}$, the estimator of f in $X = f + W$ can be written as

$$\tilde{F} = DX = \sum_{m=0}^{N-1} a_m(X_B[m])X_B[m]g_m. \tag{8.7}$$

Here, a_m is the thresholding function, which can be either hard thresholding or soft thresholding.

Hard Thresholding

A hard thresholding function is defined as follows (Mallat, 2009d):

$$a_m(x) = \begin{cases} 1 & \text{if } |x| \ge T \\ 0 & \text{if } |x| < T \end{cases}. \tag{8.8}$$

By substituting $a_m(x)$ into the estimator, we obtain the estimator with a hard thresholding function:

$$\tilde{F} = \sum_{m=0}^{N-1} \rho_T(X_B[m])g_m, \quad \text{with} \quad \rho_T(x) = a_m(x)\cdot x = \begin{cases} x & \text{if } |x| \ge T \\ 0 & \text{if } |x| < T \end{cases}. \tag{8.9}$$

The risk associated with this thresholding is

$$r_{th}(f) = r(D, f) = \sum_{m=0}^{N-1} E\left\{|f_B[m] - \rho_T(X_B[m])|^2\right\} \tag{8.10}$$

Soft Thresholding

A soft thresholding function is defined as follows (Mallat):

$$0 \le a_m(x) = \max\left(1 - \frac{T}{|x|}, 0\right) \le 1 \tag{8.11}$$

The resulting estimator $\tilde{F}$ can be written similarly to the previous case, with the thresholding function ρ_T replaced by a soft thresholding function:

$$\rho_T(x) = \begin{cases} x - T & \text{if } x \ge T \\ x + T & \text{if } x \le -T \\ 0 & \text{if } |x| \le T \end{cases} \tag{8.12}$$

Reducing the magnitude of coefficients X_B that are greater than the threshold typically results in an estimated signal $\tilde{F}$ with a smaller amplitude than the original F.

Wavelet Thresholding

Signals often contain valuable information that can be challenging to extract. The discovery of orthogonal bases and local time-frequency analysis has facilitated sparse representation of signals. An orthogonal basis is a minimal dictionary that can achieve a sparse representation if it concentrates the signal energy into a small number of vectors. The wavelet coefficients, which are few in number, provide the information we seek. These coefficients are generated by approximating the original signal as a linear combination of wavelets. For all f in $L^2(R)$,

$$P_j f = P_{j+1} f + \langle f, \psi_{j,k} \rangle \psi_{j,k}, \tag{8.13}$$

where $\langle f, \psi_{j,k} \rangle$ denotes the inner product of f and $\psi_{j,k}$, and P_j is the orthogonal projection onto V_j. In orthogonal decomposition, V_j is a subspace that satisfies $\cdots V_2 \subset V_1 \subset V_0 \subset V_{-1} \subset V_{-2} \cdots, \overline{\bigcup_{j \in \mathbb{Z}} V_j} = L^2(\mathbb{R})$ and $\bigcap_{j \in \mathbb{Z}} V_j = \{0\}$ (Daubechies, 1992).

Thresholding the wavelet coefficients preserves the local regularity of the signal. Typically, wavelet thresholding involves three steps (Shim et al., 2001; Zhang et al., 2007):

1. **Decomposition**. A filter bank of conjugate mirror filters decomposes the discrete signal into a discrete orthogonal basis. The wavelet function $\psi_{j,k}[n]$ and scale function $\phi_{j,k}[n]$ belong to the orthogonal basis

$$B = \left\{ \psi_{j,k}[n] \right\}_{L < j \leq J, 0 \leq k < 2^{-j}}, \left\{ \phi_{J,k}[n] \right\}_{0 \leq k < 2^{-J}} \tag{8.14}$$

   The scale parameter 2^j ranges from $2^L = N^{-1}$ up to $2^J < 1$, where N is the sampling rate of the signal X.
2. **Thresholding**. After decomposition, the threshold is selected. A thresholding estimator in the basis B is written as

$$\tilde{F} = \sum_{j=L+1}^{J} \sum_{k=0}^{2^{-j}} \rho_T \left(\langle X, \psi_{j,k} \rangle \right) \psi_{j,k} + \sum_{k=0}^{2^{-J}} \rho_T \left(\langle X, \phi_{J,k} \rangle \right) \phi_{J,k} \tag{8.15}$$

   where ρ_T is either a hard threshold or a soft threshold. Typically, the selected threshold is applied to all coefficients except those containing the lowest frequency energy $\langle X, \phi_{J,k} \rangle$. This approach aims to maintain the regularity of the reconstructed signal.

3. **Reconstruction**. After thresholding, all coefficients are reconstructed to form the denoised signal.

8.2 Signal Compression Using Wavelet

The bandwidth available for usage is a constraint in most of the applications. Hence, it is desirable to compress signal/data using suitable compression algorithm without compromising the quality. There are many compression algorithms. Wavelet-based compression gives high compression ratio and less distortion to the data.

8.3 Basics of Compression

Compression is the process of reducing the size of data without compromising on the quality of the data. Compression has been done for analog data in analog telephony. The technique used was called companding which stands for **Comp**ressing and exp**anding**. The method was patented by A.B.Clark of AT&T in 1928. Using this technique, signals with large dynamic range can be transmitted through narrow-band channels. Two prominent companding functions used are A-law and μ-law functions. In digital domain, compression is done mainly for transmitting signals through narrow-band channels by reducing the size. The size of the data to be stored can also be considerably reduced this way. Compression can be either lossy compression or lossless compression. The original data can be perfectly reconstructed if lossless compression is employed for compression. Lossy compression has the advantage that the compression ratio could be more compared to lossless compression. In applications like image processing where the exact reconstruction of the picture is not required, lossless compression can be deployed.

8.4 Quantization

Quantization is the process of mapping samples of an analog signal into a finite number of levels. The signal shown in Fig. 8.2 has different dynamic ranges during different time intervals. During the interval from T_1 to T_2, the dynamic range is lesser, and stepsize of the levels should be small. In contrast to this, the dynamic range of the signal during the interval from T_3 to T_4 is large, and hence, the stepsize of the levels can be made larger. This can be fulfilled by using a suitable logarithmic function for the same. A-law and μ-law functions are two such functions.

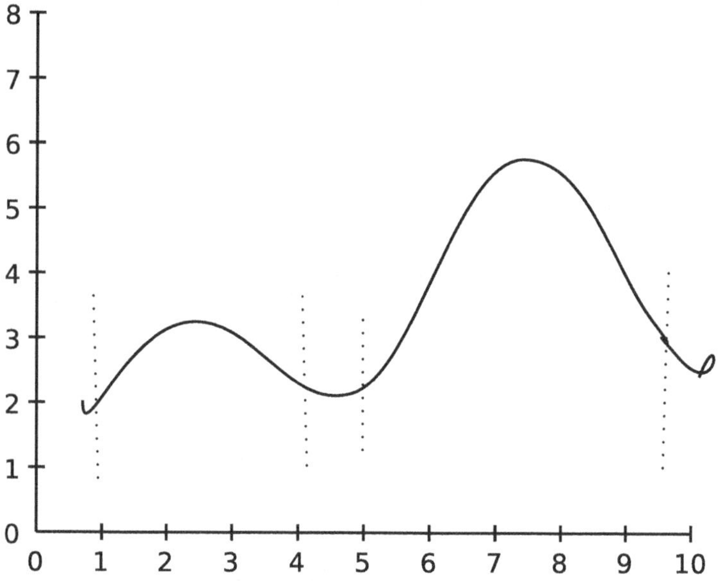

Fig. 8.2 Analog Signal

A-Law Companding

A-law companding is a function used for companding in 8-bit digital telephony. For a given x, A-law encoding is done using Eq. 8.16

$$f(x) = sgn(x) \begin{cases} \frac{A|x|}{1+log(A)}, & \text{if } |x| < \frac{1}{A}, \\ \frac{1+log(A|x|)}{1+log(A)}, & \text{if } \frac{1}{A} \leq |x| < 1 \end{cases} \tag{8.16}$$

where A is the compression parameter. A-law is adapted in Europe and the optimum value, A, used is **87.6**.

μ-Law Companding

Similar to A-law function, which was prevailing in Europe, μ-law function was used in North America and Japan. The function used for μ-law encoding is Eq. 8.17

$$f(x) = sgn(x)\frac{\log(1 + \mu|x|)}{\log(1 + \mu)} \quad \forall -1 \leq x \leq 1 \tag{8.17}$$

Fig. 8.3 Auditory mask

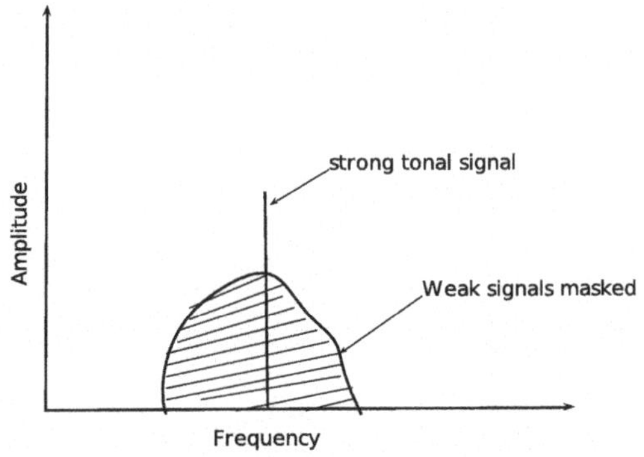

For 8-bit encoding, the value used is **255**. In comparison to A-law function, μ-law function has slightly more dynamic range. But this is with a cost of more distortion for small signals (Fig. 8.3).

8.5 Audio Compression

Compression by Encoding

The MPEG (Motion Picture Experts Group) standard is widely adopted for high-quality audio compression. This standard employs a layered compression technique. MPEG layer-I, also known as MPEG-I, was introduced in 1992 and has a bit rate of approximately 1.5 Mbps. Although MPEG-I is a lossy compression method, it achieves perceptual lossless compression by leveraging the auditory perception mechanism of the human ear.

Subjective listening tests reveal that a compression ratio of up to 6:1 can be attained by exploiting the auditory masking property of the ear. Auditory masking is a phenomenon where stronger signals mask weaker signals within the same spectral neighborhood, rendering the weaker signals inaudible. This concept of spectral masking is illustrated in Fig. 8.3. The fundamental structure of an MP3 encoder is depicted in Fig. 8.4. The process begins with the input of a PCM audio signal into a filter bank, which transforms the signal into the frequency domain. Using a psychoacoustic model, bit allocation and quantization are performed. The resulting bits are then formatted appropriately to generate the encoded signal. For decoding, as shown in Fig. 8.5, the encoded bit stream is first utilized to retrieve the frequency domain representation, which is subsequently converted back into the time domain to produce the PCM output.

The MPEG standard is divided into three layers: layer-I, layer-II, and layer-III. Layer-I is the most basic and simplest layer. Layer-II and Layer-III build on the foundation of Layer-I, providing more efficient compression at the expense of increased algorithmic complexity for encoding and decoding. Layer-I employs 32

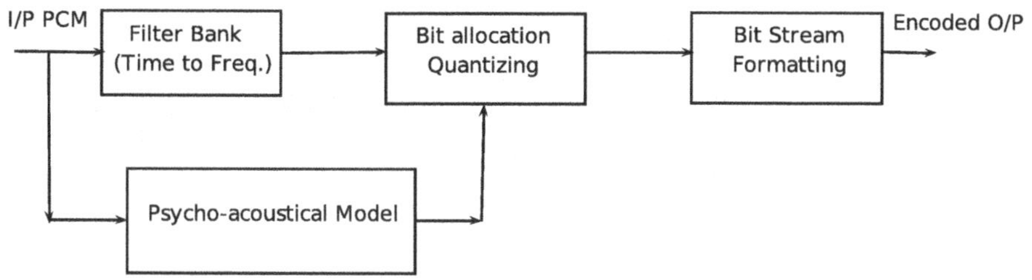

Fig. 8.4 MPEG encoder

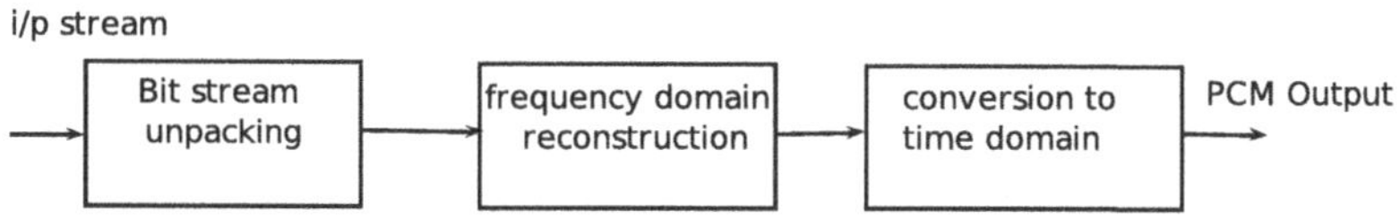

Fig. 8.5 MPEG decoder

fixed-bandwidth filter banks, where the filter bank bandwidth is selected to optimize the required time-frequency resolution for analysis. However, this layer has the drawback of not accurately representing the nonlinear characteristics of the human ear. Layer-II enhances efficiency by coding data in larger groups. Layer-III builds upon the principles of layers I and II and addresses some of their shortcomings by incorporating discrete cosine transform (DCT) encoding.

Wavelet-Based Audio Compression

Wavelet-based audio compression leverages the sparse representation of signals in the wavelet domain. Wavelets are particularly effective in concentrating both energy and perceptual information into a limited number of coefficients. As a result, applying the wavelet transform to a signal often yields many coefficients that are either zero or possess negligible values.

This compression approach is further enhanced by psychoacoustic principles. Given that human hearing is more sensitive to low frequencies and less so to high frequencies, a strategic method is employed to compress audio. This involves thresholding the detail coefficients, which represent the high-frequency components of the wavelet transform. By doing so, the introduced error remains imperceptible to the human ear.

The underlying idea of wavelet-based compression is that a regular signal can be accurately approximated using a small set of well-chosen approximation coefficients and selected detail coefficients. This approach involves discarding small-valued coefficients deemed insignificant, thus achieving data compression. The process begins with decomposing the signal into various resolutions or frequency bands using a wavelet function and computing the discrete wavelet

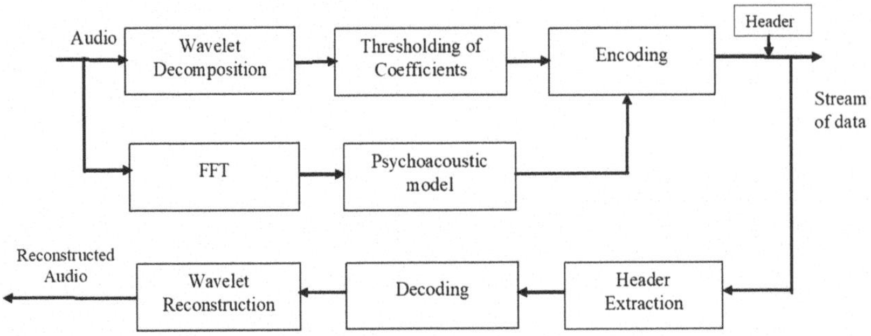

Fig. 8.6 Audio compression using wavelets

transform (DWT). Effective compression relies on the ability to represent the signal with a limited number of approximation coefficients (at an appropriate level) and some detail coefficients. After performing the wavelet transform on the audio signal, compression involves truncating coefficients that fall below a certain threshold. These small-valued coefficients can be zeroed out and subsequently used to reconstruct the signal. The compression process entails truncating insignificant coefficients and encoding the remaining ones efficiently (Fig. 8.6).

8.6 Image Compression

Image compression is a technique used to reduce the size of an image file. It is particularly useful for saving storage space and speeding up the transmission of images over the Internet. There are two main types of image compression: lossless and lossy.

Lossless Compression

In lossless compression, the original image can be perfectly reconstructed from the compressed image. This means that no image data is lost during compression. Techniques like run-length encoding (RLE), Lempel-Ziv-Welch (LZW), and Deflate are commonly used in lossless compression algorithms. Lossless compression is ideal for situations where preserving every detail of the image is essential, such as medical imaging or technical drawings. However, it typically achieves lower compression ratios compared to lossy compression.

For example, run-length encoding (RLE) is a simple and effective lossless compression technique commonly used for compressing data with long sequences of repeated values or patterns. It works by replacing consecutive identical data values with a single value and a count of how many times it repeats. RLE is particularly efficient when applied to images with large areas of uniform color or patterns. However, its effectiveness depends on the nature of the image data. In images with random or non-repeating patterns, RLE might not achieve significant compression. While RLE is relatively easy to implement and fast to execute, it may not achieve as high compression ratios as more sophisticated algorithms like those used in modern lossless image formats.

Example of Run-Length Encoding
Given the string: `AAAAABBBCAAABBBBCCAA`

The RLE encoding process for this string can be summarized as follows:

- 'AAAAA' → '5A'
- 'BBB' → '3B'
- 'C' → '1C'
- 'AAA' → '3A'
- 'BBBB' → '4B'
- 'CC' → '2C'
- 'AA' → '2A'

Thus, the RLE encoded result is: `5A3B1C3A4B2C2A`

Lossy Compression

Lossy compression reduces image file size by discarding some of the image data, which results in a reduction in image quality compared to the original. The amount of data discarded and the resulting loss of quality can be controlled by adjusting the compression parameters. Lossy compression methods exploit the limitations of human visual perception to remove redundant or less important image data. A well-known example of lossy compression is the JPEG (Joint Photographic Experts Group) format.

JPEG is designed to compress color or grayscale continuous-tone images, such as photographs, video stills, or complex graphics that resemble natural subjects. It is less effective for animations, ray tracing, line art, black-and-white documents, and typical vector graphics. While JPEG is also used for video compression, it does not specifically cater to motion video applications. The JPEG standard defines a minimal subset known as baseline JPEG, which is supported by all JPEG-compatible applications. This baseline employs the discrete cosine transform (DCT) to achieve compression:

- **Image Transformation and Downsampling**
  JPEG encodes images using any color space, though the best compression results are achieved with luminance/chrominance color spaces like YUV or YCbCr. The luminance component (Y) retains most visual information, while the chrominance components (Cb and Cr) contain less critical color information. By

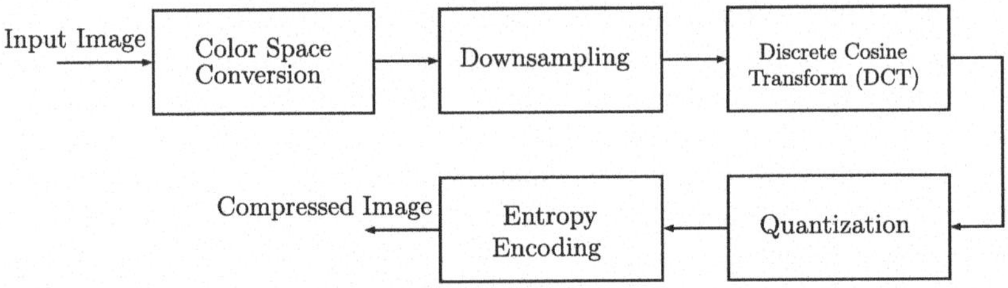

Fig. 8.7 Block diagram of JPEG Image Compression

downsampling chrominance components (e.g., reducing their resolution to 50% of the luminance resolution), JPEG leverages the human eye's reduced sensitivity to color details, achieving significant data reduction with minimal impact on perceived image quality.

- **Discrete Cosine Transform (DCT)**
  JPEG encoding uses DCT, which converts spatial image data into a frequency map. Each image is divided into 8x8 pixel blocks. The DCT separates these blocks into frequency components, with low-frequency components capturing the average values and high-frequency components capturing rapid changes in pixel values. Although the DCT step itself is lossless, it is computationally intensive.
- **Quantization**
  To reduce data size, each DCT coefficient is divided by a quantization coefficient and rounded to an integer. Higher quantization coefficients result in greater data loss. Quantization is tailored separately for luminance and chrominance components, with chrominance data typically quantized more heavily. The quantization process is controlled by the "quality" setting of the JPEG compressor, affecting the final image quality and file size.
- **Encoding**
  After quantization, the coefficients are compressed using Huffman coding or, optionally, arithmetic encoding. This step removes redundant data and prepares the JPEG data for transmission or storage in an image file format (Fig. 8.7).

8.7 JPEG 2000

JPEG 2000 (JP2) is an image compression standard and coding system created by the Joint Photographic Experts Group committee in 2000. It was designed to supersede the original discrete cosine transform-based JPEG standard from 1992 with a newly designed, wavelet-based method.

JPEG 2000 code streams are regions of interest that offer several mechanisms to support spatial random access or region of interest access at varying degrees of

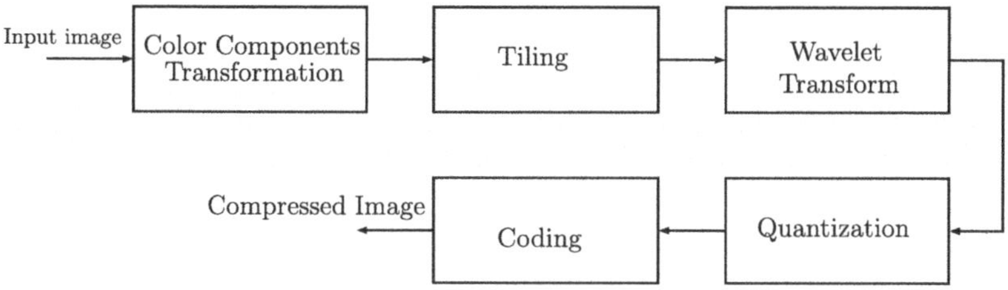

Fig. 8.8 JPEG 2000

granularity. It is possible to store different parts of the same picture using different quality levels. The code stream obtained after compression of an image with JPEG 2000 is scalable in nature. By truncating the code stream at any point, one may obtain a representation of the image at a lower resolution or signal-to-noise ratio, thus achieving scalable compression.

The simplified structures of the encoder and decoder of JPEG 2000 are shown in Fig. 8.8. For a multiple-component image, the major processing steps of the encoder are:

- Color components transformation
- Tiling
- Wavelet transform
- Quantization
- Coding

The decoder reverses the steps performed by the encoder, except for the rate-distortion optimization step.

Color Components Transformation

Initially, images are transformed from the RGB color space to another color space, leading to three components that are handled separately. There are two possible choices:

Irreversible Color Transform (ICT) uses the well-known YCBCR color space. It is called "irreversible" because it must be implemented in floating or fixed point and causes round-off errors.

Reversible Color Transform (RCT) uses a modified YUV color space that does not introduce quantization errors, making it fully reversible. Proper implementation of the RCT requires rounding as specified, which cannot be expressed exactly in matrix form.

Tiling

After color transformation, the image is split into rectangular regions called tiles, which are transformed and encoded separately. Tiles can be of any size, and it is possible to consider the whole image as a single tile. Once the size is chosen, all tiles will have the same size, except optionally those on the right and bottom borders.

Wavelet Transform

These tiles are then wavelet transformed into an arbitrary depth, in contrast to JPEG 1992 which uses an 8×8 block-size discrete cosine transform. JPEG 2000 uses two different wavelet transforms:

- **Irreversible:** The CDF 9/7 wavelet transform. It is called "irreversible" because it introduces quantization noise that depends on the precision of the decoder.
- **Reversible:** A rounded version of the biorthogonal CDF 5/3 wavelet transform. It uses only integer coefficients, so the output does not require rounding (quantization) and does not introduce any quantization noise. It is used in lossless coding. The wavelet transforms are implemented by the lifting scheme or by convolution.

Quantization

After the wavelet transform, the coefficients are scalar-quantized to reduce the number of bits needed to represent them, at the expense of quality. The output is a set of integer numbers which have to be encoded bit-by-bit. The parameter that can be changed to set the final quality is the quantization step: the greater the step, the greater the compression and the loss of quality. With a quantization step that equals 1, no quantization is performed (used in lossless compression).

Coding

The quantized subbands are split further into precincts, which are rectangular regions in the wavelet domain. They are typically selected in a way that the coefficients within them across the subbands form approximately spatial blocks in the (reconstructed) image domain. Precincts are split further into code blocks, which are located in a single subband and have equal sizes, except those located at the edges of the image. The encoder encodes the bits of all quantized coefficients of a code block, starting with the most significant bits and progressing to less significant bits through the embedded block coding with optimal truncation (EBCOT) scheme. In this encoding process, each bit plane of the code block is encoded in three coding passes: first encoding bits (and signs) of insignificant coefficients with significant neighbors, then refining bits of significant coefficients, and finally coefficients without significant neighbors.

8.8 Video Compression

Video compression involves reducing the file size of a video by discarding some information or quality. Digital video is a crucial visual source in contemporary media. A device or program that compresses a signal is known as an encoder, while a device or program that decompresses a signal is a decoder. Together, these components are referred to as a CODEC. The information in digital video is highly correlated and redundant. The primary objective of a lossy video CODEC is to reduce this redundancy by removing information that is not perceivable by the human visual system.

Typically, video compression schemes start with a full-color video signal in the RGB color space. RGB space represents pixels with three numbers indicating the relative proportions of red, green, and blue. RGB systems usually treat all three color components with equal precision, meaning all colors are considered equally important. However, luminance is present in all three color components, creating redundancy. The YCrCb color space is more efficient as it separates luminance (Y) from color (Cr and Cb) components. Since the human visual system is more sensitive to luminance than color, some color information can be removed without significantly affecting visual quality. In 4:2:2 YCrCb systems, downsampling is applied only in the horizontal direction, achieving a 3:2 compression ratio. In 4:1:1 YCrCb conversion, Cr and Cb signals are downsampled both horizontally and vertically, resulting in a 2:1 compression ratio.

In addition to spectral redundancy, image and video signals contain spatial information that we cannot always perceive. Compacting this information without significantly affecting visual quality is challenging. One approach is to transform the image into a different representation. A good transformation should represent the signal with a few significant values or compact the image's energy. The discrete cosine transform (DCT) and discrete wavelet transform (DWT) are the most widely used methods for this purpose.

Spatial compression alone does not suffice for achieving typical compression goals. Since video signals are sampled at rates of 15–30 frames per second, subsequent frames in a video sequence often appear nearly identical. Only small portions of the scene change from frame to frame, with background information remaining similar if camera movement and illumination changes are minimal. This creates temporal redundancy between frames. Modern video coders address this by employing motion estimation and motion compensation (ME/MC) techniques. Alternatively, a three-dimensional transform, applied in both time and spatial

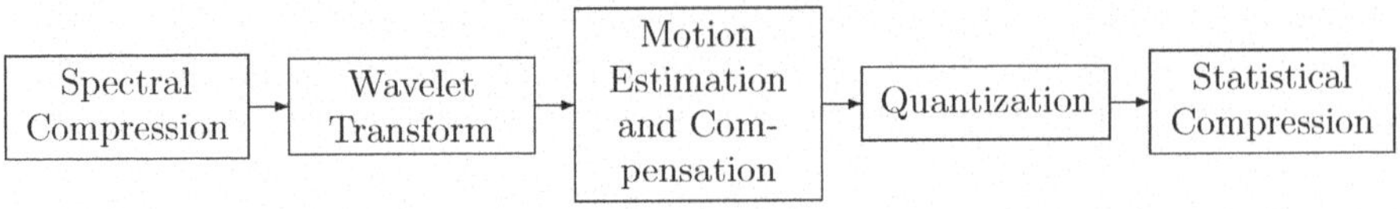

Fig. 8.9 Block diagram of wavelet-based video coder

domains, can be used instead of a two dimensional transform combined with ME/MC.

Even with redundancy reduction techniques, some redundancy remains in the compressed signal due to the statistical properties of video signals. To further reduce these redundancies, statistical lossless coding schemes are employed. Examples include run-length coding, Huffman coding, and arithmetic coding.

Wavelet-Based Video Compression

Currently, there are no standardized video CODECs that exclusively use wavelet-based transform coding, though some implementations exist. Early applications of wavelets in video were seen in the IMMIX Cube NLE systems, which used wavelets to achieve better video quality at lower bit rates compared to Motion-JPEG. Despite IMMIX's challenges, wavelets are employed in Motion-JPEG2000, the video extension of JPEG-2000, as well as in Indeo v4 and v5 (Fig. 8.9).

The wavelet-based video coder involves several key processing steps designed to efficiently compress video data:

Spectral Compression This step reduces the redundancy in the video signal by transforming it into a more compact representation, often using techniques such as the discrete cosine transform (DCT) or other spectral methods.

Wavelet Transform The video signal is transformed using wavelets, which provide a multiresolution analysis of the data. This helps in capturing both high- and low-frequency components of the video, leading to better compression efficiency.

Motion Estimation and Compensation This process involves analyzing and compensating for movements between frames. By estimating the motion between successive frames, the coder can reduce temporal redundancy, focusing only on the differences or changes between frames.

Quantization The transformed coefficients are quantized to reduce their precision. This step involves approximating the coefficients to a smaller set of values, which reduces the amount of data needed to represent them.

Statistical Compression Finally, statistical compression techniques, such as Huffman coding or Arithmetic coding, are applied to the quantized data to further compress it by encoding the data more efficiently based on its statistical properties.

The biorthogonal Cohen–Daubechies–Feauveau 9/7 filter, known as the "9/7 filter," is favored in image and video coding for its excellent energy compaction properties. The lifting scheme, introduced by Wim Sweldens in the mid-1990s, provides a framework for designing biorthogonal wavelet filters usable in various discrete wavelet transforms. Unlike classical constructions based on the Fourier transform, the lifting scheme enables the creation of second-generation wavelets, which are not merely translates and dilates of the mother wavelet. These second-generation wavelets can be applied in areas where the Fourier transform is not available, such as data segmentation, curve or surface analysis, and irregularly sampled data.

The primary advantage of the lifting scheme in wavelet-based video CODECs is its reduction of computational complexity. It allows for faster implementation and fully in-place calculation of the wavelet transform, eliminating the need for auxiliary memory to store temporary data. The lifting scheme can reduce computation complexity by half compared to the standard iterated FIR filter bank algorithm (classical DWT). This transform is more efficient than the fast Fourier transform (FFT), with the lifting scheme improving speed by another factor due to optimal utilization of similarities between odd and even values or high- and low-pass filters.

For classical wavelet transforms, the Fourier transform is used to ensure exact reconstruction of the original signal. However, with the lifting scheme, it is easier to reconstruct the original signal by undoing the operations of the forward transform. Complex wavelets for motion estimation showed potential for approximate shift invariance and good directional selectivity. Despite initial challenges with perfect reconstruction and frequency characteristics using short support complex FIR filters in a single tree, advancements by Kingsbury and others led to the development of the dual-tree complex wavelet transform (DT CWT).

The dual filter tree implementation of the DT CWT aims to achieve approximate shift invariance by doubling the sampling rate at each level of the tree, with even spacing of samples. This implementation involves two trees of real filters, producing real and imaginary parts of the complex coefficients. Successful operation depends on differences between the filters in the two trees, with specific sampling and delay differences required to achieve optimal shift invariance. To invert the transform, perfect reconstruction filters are applied to invert each tree separately and then average the results.

8.9 Solved Problems

Problem Write a MATLAB program to generate a signal, add noise, and denoise with hard thresholding.

Solution MATLAB program:

```
fs = 1000;              % Sampling frequency
    t = 0:1/fs:1;           % Time vector
    f = 5;                  % Frequency
    original_signal = sin(2*pi*f*t);

    % Add Gaussian noise
    noise_amplitude = 0.2;
    noisy_signal = original_signal + noise_amplitude * randn(size(original_signal
        ));

    % Display the original and noisy signals
    figure;
    subplot(3,1,1);
    plot(t, original_signal);
    title('Original Signal');

    subplot(3,1,2);
    plot(t, noisy_signal);
    title('Noisy Signal');

    % Perform wavelet decomposition
    wavelet_type = 'db4';
    level = 6;
    [C, L] = wavedec(noisy_signal, level, wavelet_type);

    % Threshold the wavelet coefficients (hard thresholding)
    threshold = 1.0;
    C_thresholded = wthresh(C, 'h', threshold);

    % Reconstruct the denoised signal
    denoised_signal = waverec(C_thresholded, L, wavelet_type);

    % Display the denoised signal
    subplot(3,1,3);
    plot(t, denoised_signal);
    title('Denoised Signal');

    % Add labels and legend
    xlabel('Time (s)');
    legend('Original', 'Noisy', 'Denoised');

    % Display legend
    legend('Location', 'Best');
```

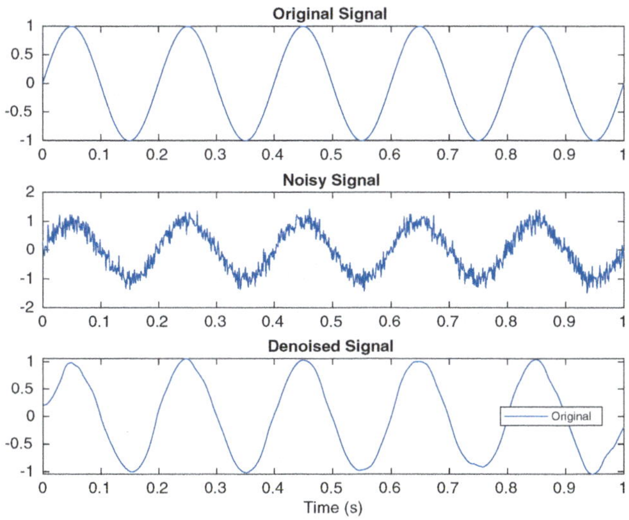

Fig. 8.10 Denoising using hard thresholding

See Fig. 8.10.

Problem Write a MATLAB program to generate a signal, add noise, and denoise with soft thresholding.

Solution MATLAB program:

```
fs = 1000;             % Sampling frequency
    t = 0:1/fs:1;          % Time vector
    f = 5;                 % Frequency
    original_signal = sin(2*pi*f*t);

    % Add Gaussian noise
    noise_amplitude = 0.2;
    noisy_signal = original_signal + noise_amplitude * randn(size(original_signal
        ));

    % Display the original and noisy signals
    figure;
    subplot(3,1,1);
    plot(t, original_signal);
    title('Original Signal');

    subplot(3,1,2);
    plot(t, noisy_signal);
    title('Noisy Signal');

    % Perform wavelet decomposition
    wavelet_type = 'db4';
    level = 6;
```

```
        [C, L] = wavedec(noisy_signal, level, wavelet_type);

        % Threshold the wavelet coefficients (soft thresholding)
        threshold = 1.0;
        C_thresholded = wthresh(C, 's', threshold);

        % Reconstruct the denoised signal
        denoised_signal = waverec(C_thresholded, L, wavelet_type);

        % Display the denoised signal
        subplot(3,1,3);
        plot(t, denoised_signal);
        title('Denoised Signal');

        % Add labels and legend
        xlabel('Time (s)');
        legend('Original', 'Noisy', 'Denoised');

        % Display legend
        legend('Location', 'Best');
```

See Fig. 8.11.

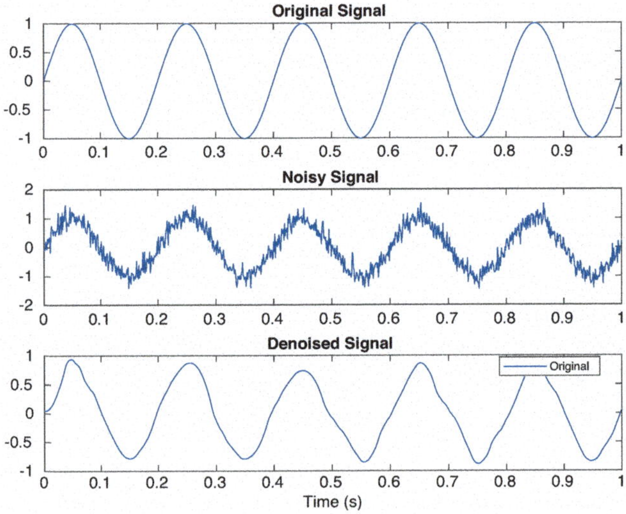

Fig. 8.11 Denoising using soft thresholding

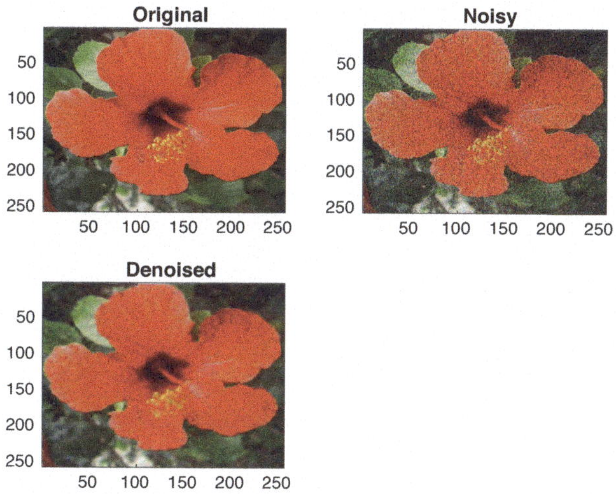

Fig. 8.12 Denoising of colour image (Figure courtesy: Matlab Wavelet Tool box)

Problem Write a MATLAB program to denoise a color image.

Solution MATLAB Program:

```
load colorflower
    subplot(2,2,1)
    imagesc(colorflower.Orig)
    title('Original')
    subplot(2,2,2)
    imagesc(colorflower.Noisy)
    title('Noisy')
    imden = wdenoise2(colorflower.Noisy, 2, 'CycleSpinning', 1);
    subplot(2,2,3)
    imagesc(imden)
    title('Denoised')
```

See Fig. 8.12.

Chapter 9
Optimization Methods for Wavelet Design

9.1 Optimization Methods

Optimization is defined as the task of searching for a set of decision variables which would minimize or maximize an objective function subjected to specific constraints. The objective function is simply the value that we are trying to optimize. It can be either to maximize or minimize a function. The variables of the objective function that the optimizer can modify to obtain an optimal solution are called decision variables. Finally, we put some constraints on the decision variables of the problem in order to control the range of each variable.

Consider the following example in which our objective is to maximize the function $f(x, y) = x^2 + y^2 + C$, where the decision variables are (x, y). The objective function is to maximize $f(x, y)$. The optimization algorithm starts with a random point, which we define as the initial guess P_1. The objective function corresponding to it is P_1'. Using a local idea or some rule, we find the point P_2 in the direction where the objective function is improved. This point P_2 has a better objective function value than P_1, such that P_2 is better than the initial guess. The process is repeated, and finally, we reach the optimal point P. The process is illustrated in Fig. 9.1. The whole process can be defined as a search for the optimal values of x and y such that the function $f(x, y)$ is maximized.

In general, a multivariable single-objective optimization problem can be defined as

$$\textbf{Minimize}: f(\mathbf{x})$$

$$\textbf{Subject to}:$$

$$g_j(\mathbf{x}) \geq 0 \quad \text{for } j = 1, 2, 3, \ldots, J$$

$$h_k(\mathbf{x}) = 0 \quad \text{for } k = 1, 2, 3, \ldots, K$$

$$x_i^{(L)} \leq x_i \leq x_i^{(U)} \quad \text{for } i = 1, 2, \ldots, N$$

M. S. Sinith et al., *Wavelets and Multi-Resolution Analysis*,
https://doi.org/10.1007/978-3-031-92750-8_9

Fig. 9.1 Optimization of $f(x, y) = x^2 + y^2 + C$

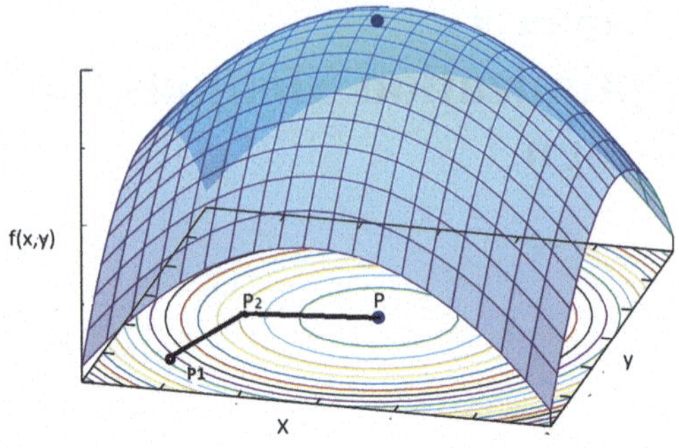

- Where x is the decision variable vector.
- $f(x)$ is the objective function, and here the objective function is a minimization function.
- $g_j(x)$ is the jth inequality constraint.
- $h_k(x)$ is the kth equality constraint.
- $x_i^{(L)}$ and $x_i^{(U)}$ are the lower bound and upper bound of the ith decision variable.

Steps in Optimization Task

1. **Define the Objective Function:** Identify the quantity to be optimized, such as cost, performance, or efficiency, and express it as the objective function $f(\mathbf{x})$. The function represents the relationship between the decision variables and the value to be optimized.
2. **Identify Decision Variables:** Determine the parameters, represented as $\mathbf{x} = [x_1, x_2, \ldots, x_N]$, that influence the objective function. These are the variables that the optimization process will adjust to find the optimal solution.
3. **Specify Constraints:** Define any constraints that the solution must satisfy:
   - **Inequality Constraints:** Conditions such as $g_j(\mathbf{x}) \geq 0$, which restrict the solution space to ensure that certain criteria are met.
   - **Equality Constraints:** Conditions such as $h_k(\mathbf{x}) = 0$, which impose exact requirements on the solution.
   - **Variable Bounds:** Define the permissible range for each decision variable using $x_i^{(L)} \leq x_i \leq x_i^{(U)}$. These bounds reflect physical or practical limitations.
4. **Choose the Optimization Objective:** Decide whether the goal is to minimize or maximize the objective function. For maximization problems, the objective function can be transformed into a minimization problem by minimizing $-f(\mathbf{x})$.
5. **Explore the Feasible Region:** Systematically search within the feasible region defined by the constraints to find the values of the decision variables that optimize

the objective function. This involves evaluating $f(\mathbf{x})$ at different points within the permissible range.

6. **Identify the Optimal Solution:** Determine the set of decision variables $\mathbf{x}$ that satisfies all constraints and achieves the best possible value of the objective function, ensuring both optimality and feasibility.

9.2 Properties of Practical Optimization Problems

Non-differentiable Functions and Constraints

It is well known that non-differentiable functions cannot be used for gradient-based optimization because the optimization algorithms assume a smooth change in gradients. In order to handle non-differentiable functions in optimization, we need to approximate the non-differentiable function using a smooth differentiable function. However, the approximation process makes the function highly nonlinear, which makes the optimization difficult to handle, as illustrated in Fig. 9.2.

The function $y = |x|$ is **piecewise linear**, not nonlinear. It consists of two linear segments:

$$y = \begin{cases} x, & x \geq 0 \\ -x, & x < 0 \end{cases} \tag{9.1}$$

Although each segment is linear, the absolute value function is not differentiable at $x = 0$ because the left-hand and right-hand derivatives are different. This lack of smoothness at $x = 0$ is what makes gradient-based optimization problematic.

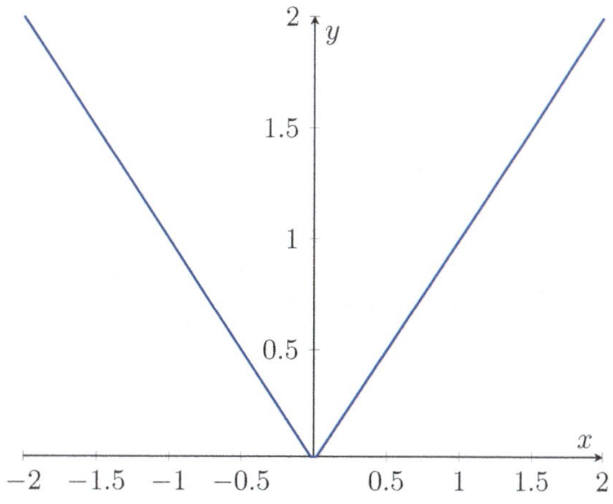

Fig. 9.2 Plot of the non-differentiable function $y = |x|$

Discontinuous Function

If the function is discontinuous, the optimization algorithm may get stuck at one portion and may not reach the global minimum.

Mixed Variables

In some cases, variables might have different types; some might be real, discrete (choice), binary, or integer-valued. For those cases, different evolutionary operators must be applied to different types of variables.

Multimodal Functions

These kinds of functions are challenging in an optimization task because there exist many optimal solutions. Considering the figure, for a function $f(x)$, we have optimal solutions at x_1, x_2, and x_3. The optimization algorithm will give any of these points as an optimal solution. However, other points may also be useful in the design problem and for making strategic decisions. Therefore, we cannot discard these solutions. This poses a challenge in running optimization algorithms for multimodal functions.

Multiple Objectives

A multi-objective optimization problem is an optimization problem that involves multiple objective functions. In mathematical terms, a multi-objective optimization problem can be formulated as

$$\text{minimize} : (f_1(x), f_2(x), f_3(x), \ldots, f_k(x)) \quad \text{where} \quad k \geq 2$$

In multi-objective optimization, there does not typically exist a feasible solution that minimizes all objective functions simultaneously. Therefore, attention is paid to Pareto optimal solutions, that is, solutions that cannot be improved in any of the objectives without degrading at least one of the other objectives.

Evolutionary Computation

Evolutionary computation is a subfield of computational intelligence that uses ideas and gets inspiration from natural evolution. These algorithms are inspired by biological evolution of human beings, social behavior of insects, etc. They can be used to solve complex optimization problems by generating, evaluating, and modifying a population-based solution. Evolutionary algorithms can be either nature-inspired or based on natural evolution and genetics. Nature-inspired algorithms mimic biological phenomena or processes. Algorithms based on natural evolution create offspring (new solutions) based on variation operators such as mutation and recombination. The solutions which are closer to the optimum solutions are retained and others are discarded, similar to survival of the fittest in evolution. EC techniques are population-based algorithms, meaning multiple solutions exist.

The common underlying idea behind all these techniques is the same: Given a population of individuals, the environmental pressure causes natural selection (survival of the fittest), and this leads to a rise in the fitness of the population. A fitness function is defined as the objective function. Based on this fitness, some of the better candidates are chosen to seed the next generation by applying recombination and/or mutation to them. Recombination is an operator applied to two or more selected candidates (the so-called parents) and results in one or more new candidates (the children). Mutation is applied to one candidate and results in one new candidate. Executing recombination and mutation leads to a set of new candidates (the offspring) that compete—based on their fitness with the old ones for a place in the next generation. This process can be iterated until a candidate with sufficient quality (a solution) is found or a previously set computational limit is reached.

In this process, there are two fundamental forces that form the basis of evolutionary systems:

- Variation operators (recombination and mutation) create the necessary diversity and thereby facilitate novelty, while
- Selection acts as a force pushing quality.

9.3 Genetic Algorithm

Genetic algorithm (GA) is a search-based optimization technique based on the principles of genetics and natural selection. GAs were developed by John Holland and his students and colleagues at the University of Michigan. In GA, we have a set of possible solutions to the given problem. These solutions then undergo recombination and mutation (like in natural genetics), producing new children, and the process is repeated over various generations. Each individual (or candidate solution) is assigned a fitness value (based on its objective function value), and

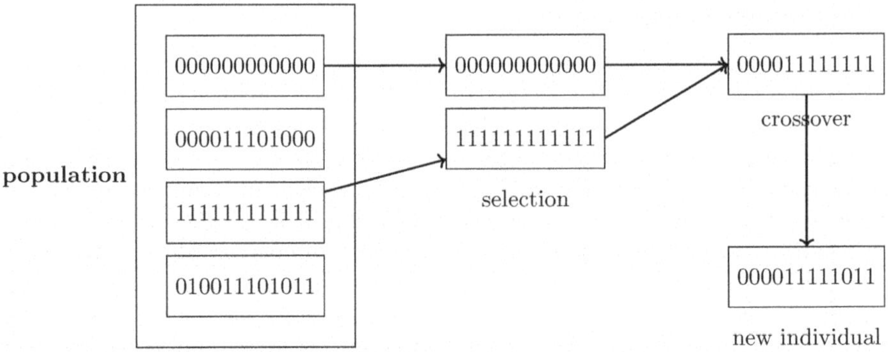

Fig. 9.3 Basic operators in evolutionary algorithms

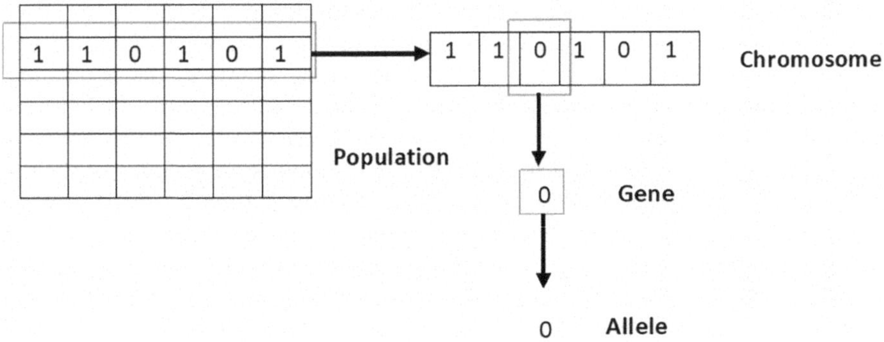

Fig. 9.4 Basic terminologies in GA

the fitter individuals are given a higher chance to mate and yield more "fitter" individuals. This is in line with the Darwinian Theory of "Survival of the Fittest." In this way, we keep "evolving" better individuals or solutions over generations until we reach a stopping criterion (Fig. 9.3).

Basic Terminologies

- **Population**: It is a subset of all the possible solutions to the given problem.
- **Chromosomes**: A chromosome is one such solution to the given problem.
- **Gene**: A gene is one element position of a chromosome.
- **Allele**: It is the value a gene takes for a particular chromosome (Fig. 9.4).
- **Genotype**: Genotype is the population in the computation space. In the computation space, the solutions are represented in a way which can be easily understood and manipulated using a computing system.

- **Phenotype**: Phenotype is the population in the actual real-world solution space in which solutions are represented in a way they are represented in real-world situations.
- **Decoding and Encoding**: For simple problems, the phenotype and genotype spaces are the same. However, in most of the cases, the phenotype and genotype spaces are different. Decoding is a process of transforming a solution from the genotype to the phenotype space, while encoding is a process of transforming from the phenotype to genotype space. Decoding should be fast as it is carried out repeatedly in a GA during the fitness value calculation.
- **Fitness Function**: A fitness function simply defined is a function which takes the solution as input and produces the suitability of the solution as the output. In some cases, the fitness function and the objective function may be the same, while in others it might be different based on the problem.
- **Genetic Operators**: These alter the genetic composition of the offspring. These include crossover, mutation, selection, etc.

Basic Algorithm

The basic structure of a genetic algorithm (GA) is as follows:

We start with an initial population (which may be generated at random or seeded by other heuristics) and select parents from this population for mating. Apply crossover and mutation operators on the parents to generate new offspring. Finally, these offspring replace the existing individuals in the population, and the process repeats. In this way, genetic algorithms actually try to mimic human evolution to some extent (Fig. 9.5).

Selection The process that determines which solutions are to be preserved and allowed to reproduce and which ones deserve to die out. The primary objective of the selection operator is to emphasize the good solutions and eliminate the bad solutions in a population while keeping the population size constant. There are different techniques to implement selection in genetic algorithms. They are: tournament selection, Roulette wheel selection, proportionate selection, rank selection, steady-state selection, etc.

Crossover The crossover operator is used to create new solutions from the existing solutions available in the mating pool after applying selection operator. This operator exchanges the gene information between the solutions in the mating pool. The most popular crossover selects any two solutions strings randomly from the mating pool, and some portion of the strings is exchanged between the strings. The selection point is selected randomly. A probability of crossover is also introduced in order to give freedom to an individual solution string to determine whether the solution would go for crossover or not.

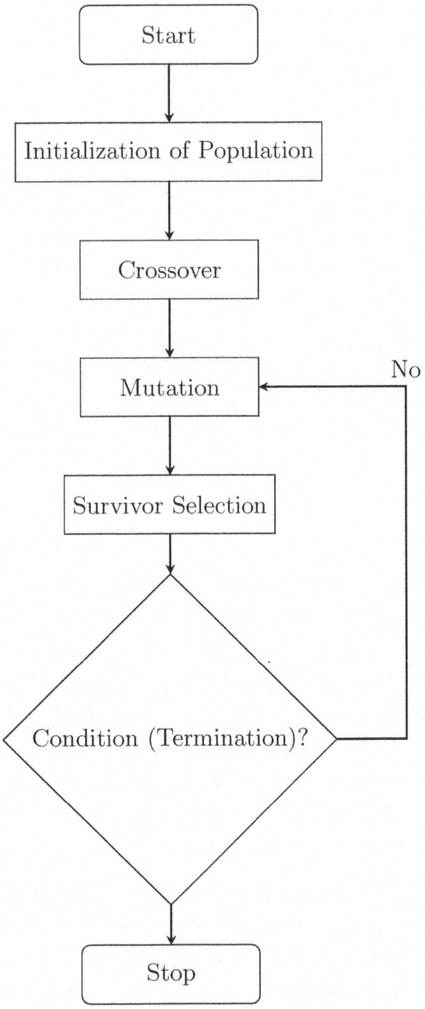

Fig. 9.5 Flow chart for genetic algorithm

Mutation Mutation is the occasional introduction of new features into the solution strings of the population pool to maintain diversity in the population. It is usually applied with a low probability P_m. If the probability is very high, the GA gets reduced to a random search. Mutation is the part of the GA which is related to the "exploration" of the search space. It has been observed that mutation is essential to the convergence of the GA, while crossover is not. The major mutation operators are: bit flip mutation, swap mutation, scramble mutation, inversion mutation, etc.

9.4 Differential Evolution

Differential evolution (DE) was proposed by Rainer Storn and Kenneth Price in 1995 (not 1976). It is a population-based stochastic evolutionary computation technique. It inherits the parallel search space property for escaping local minima. While searching for the optimal solution, some solutions may get trapped in local minima, but other solutions may help in escaping these local minima. DE was proposed by the authors for minimizing non-differentiable and nonlinear continuous space functions. DE requires few control parameters, is robust, easy to use, and allows function evaluation to be done in parallel.

Basic Principle
Differential evolution is a stochastic, population-based technique. Each solution is referred to as a genome or chromosome. Each chromosome undergoes mutation followed by recombination (crossover). There are several terminologies we need to consider (Fig. 9.6):

- **Target vector:** The solution that is undergoing evolution. It is the vector which is being perturbed.
- **Mutant vector:** Also known as the donor vector. It is the vector generated by adding the weighted difference between two randomly chosen vectors to the third vector in the population, i.e., the vector produced after mutation.
- **Trial vector:** A vector produced after recombination. It is produced by mixing variables of the target vector and the mutant vector.

A greedy selection is performed between target and trial vectors. Whichever solution is better will be selected to survive to the next generation. The selection of the better solution is performed after the generation of all trial vectors (Fig. 9.7).

Initialization
DE is a population-based optimization technique that begins with an initial population, which can either be generated randomly or seeded by other heuristics. We define upper and lower bounds for each parameter: $x_j^L \leq x_{j,i,1} \leq x_j^U$. The initial parameter values are randomly selected uniformly from the interval $[x_j^L, x_j^U]$.

Mutation
Mutation expands the search space. For each target vector, choose three other vectors randomly. The mutant vector is calculated by adding the difference between two vectors to the third. For a given parameter x_i, G, where G is the generation,

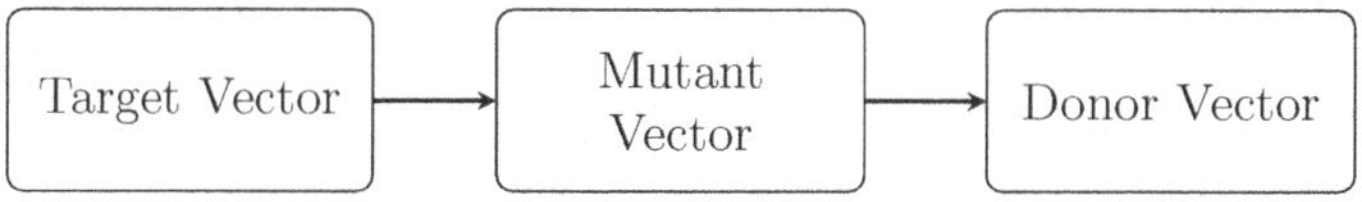

Fig. 9.6 Basic steps of differential evolution

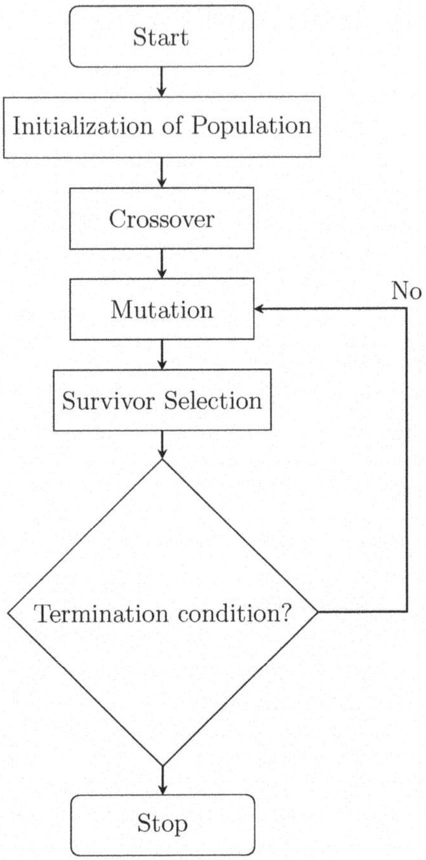

Fig. 9.7 Flow chart for differential evolution

randomly select three vectors $x_{r1,G}, x_{r2,G}, x_{r3,G}$ such that i, r_1, r_2, r_3 are distinct. The mutant vector is then given by

$$v_{i,G+1} = x_{r1,G} + F(x_{r2,G} - x_{r3,G})$$

where F is a constant mutation factor from [0, 2] and $v_{i,G+1}$ is the mutant vector.

Crossover

Recombination (or crossover) incorporates successful solutions from the previous generation. To balance the differential mutation search strategy, DE applies uniform crossover to construct trial vectors. A trial vector is constructed from values that have been copied from two diverse vectors. Elements of the donor vector enter the trial vector with probability C_r. In particular, DE crosses each vector as follows:

$$u_{i,j}^G = \begin{cases} v_{i,j}^G & \text{if } \mathrm{rand}_j \le C_r \text{ or } j = j_{rand} \\ x_{i,j}^G & \text{otherwise} \end{cases}$$

where:

- $u_{i,j}^G$ is the jth component of the trial vector for the ith individual at generation G.
- $v_{i,j}^G$ is the jth component of the mutant vector for the ith individual at generation G.
- $x_{i,j}^G$ is the jth component of the target vector for the ith individual at generation G.
- C_r is the crossover probability.
- rand_j is a random number between 0 and 1.
- j_{rand} is a randomly selected index to ensure that at least one component of the trial vector is taken from the mutant vector.

Selection
In this stage, we determine if the trial vector has an objective function value that is less than or equal to that of its target vector. If this is the case, DE swaps the target vector in the next iteration; otherwise, the target retains its place in the population. This process is carried out by comparing each trial vector with the target vector from which the parameters are cloned. After the population is updated, mutation, recombination, and selection are repeated until the optimum value is found or a predefined stop criterion is reached (e.g., a certain number of iterations).

9.5 Particle Swarm Optimization

Particle swarm optimization (PSO) algorithm is a stochastic optimization technique based on swarm, which was proposed by Eberhart and Kennedy (1995) and Kennedy and Eberhart (1995). PSO algorithm simulates animal's social behavior, including insects, herds, birds, and fishes. These swarms conform a cooperative way to find food, and each member in the swarms keeps changing the search pattern according to the learning experiences of its own and other members. Main design idea of the PSO algorithm is closely related to two researches: One is evolutionary algorithm, just like evolutionary algorithm; PSO also uses a swarm mode which makes it to simultaneously search large region in the solution space of the optimized objective function. The other is artificial life, namely it studies the artificial systems with life characteristics.

Basic Principle
The main goal of the PSO algorithm is to search in the search space for the positions/locations that are close to the global minimum or maximum solution(s). The dimension of the search space is determined by the number of parameters that need to be optimized. For example, if the search space has n dimensions, then the number of variables in the objective function is also n. The PSO algorithm has many parameters. The current position of each particle is used to calculate the fitness value at that location. Each particle has three parameters, namely, position ($x^i \in \mathbb{R}$), velocity, i.e., rate of position change, (v^i), and the previous best position (p_{best}^i).

In addition, the position of the particle that has the best fitness value is called the global best position, and it is denoted by g_{best}.

The position of each particle is denoted by x_i, and it is represented by a set of coordinates that represents a point in the search space. During the searching process, the current positions of all particles are evaluated using the fitness function. The fitness value of each particle is then compared with the current position, and the best position is stored in the previous best positions (p^i_{best}). In other words, the previous best positions store the positions of the particles that have better values. After comparing with fitness value position, it is updated as

$$x^i(t+1) = x^i(t) + v^i(t+1) \tag{9.2}$$

where particles' velocities are updated as

$$v^i(t+1) = wv_i(t) + C_1r_1(P^i_{best} - x^i(t)) + C_2r_2(g_{best} - x^i(t)) \tag{9.3}$$

where w is the inertia weight, C_1 represents the cognition learning factor, C_2 indicates the social learning factors, r_1, r_2 are the uniformly generated random numbers in the range of [0, 1], and (p^i_{best}) is the best solution of the ith particle. Since the velocity of the particles depends on random variables, the particles move randomly. Thus, the motion of the particles is called random walk (Fig. 9.8).

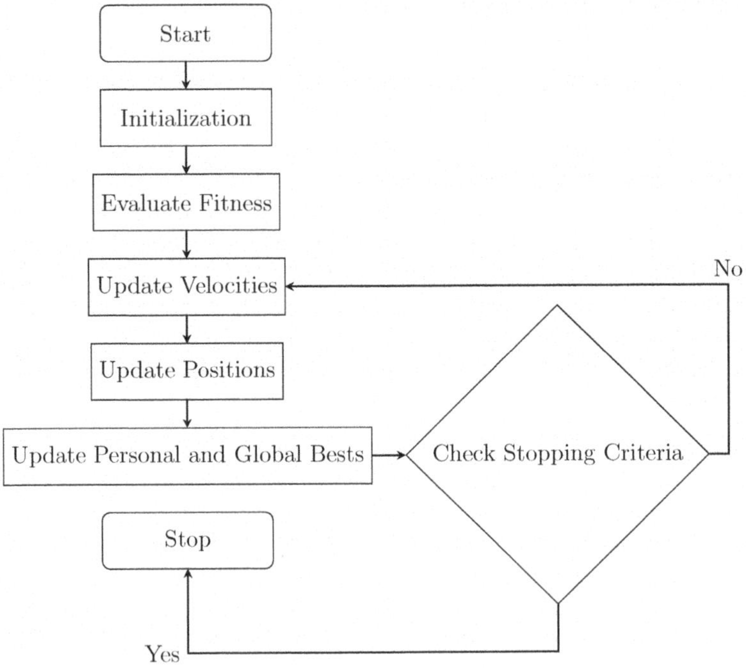

Fig. 9.8 Flow chart for particle swarm optimization

9.6 Application

Evolutionary algorithms (EAs) have emerged as a powerful tool for designing and optimizing wavelets tailored to specific signal processing applications. Unlike traditional analytical methods, which often rely on predefined mathematical structures, EAs enable the adaptive evolution of wavelet functions by iteratively improving filter coefficients or wavelet parameters based on a defined fitness function. This approach is particularly useful for applications where conventional wavelets may not be optimal, such as speech and music signal analysis, biomedical signal processing, and image compression.

SSM Wavelets for Musical Instruments

The time domain representation of a single note played by a musical instrument typically exhibits a repeating element. This repeating element can be represented as the scaling function of that instrument. Figures 9.9 and 9.10 illustrate the waveforms and their corresponding fast Fourier transforms (FFTs) for a single note played by a flute and a violin. From the spectral plots, it is evident that the flute signal has one fundamental frequency and two dominant harmonics, whereas the violin signal has one fundamental frequency and more than seven harmonics.

In higher-dimensional spaces, the ordinary particle swarm optimization (PSO) algorithm did not converge to exact values with zero error. Consequently, a modification was made to the standard PSO algorithm. Instead of a single initialization followed by iterations, an arbitrary point is first assigned to each particle in the hyperplane. These particles are then allowed to move in the N-dimensional plane. After a sufficient number of iterations (200), these particles converge to a point

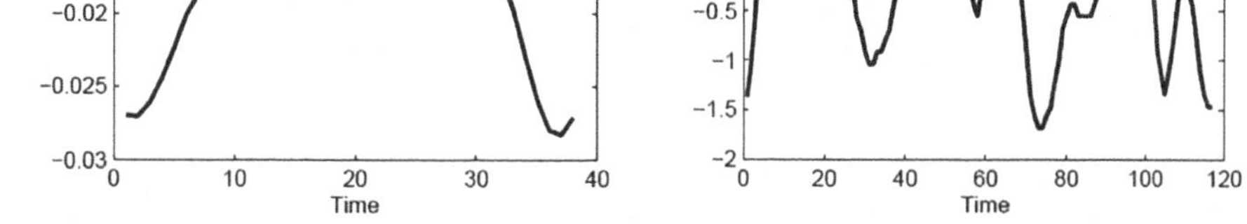

Fig. 9.9 Waveforms of a single note played by a flute and a violin

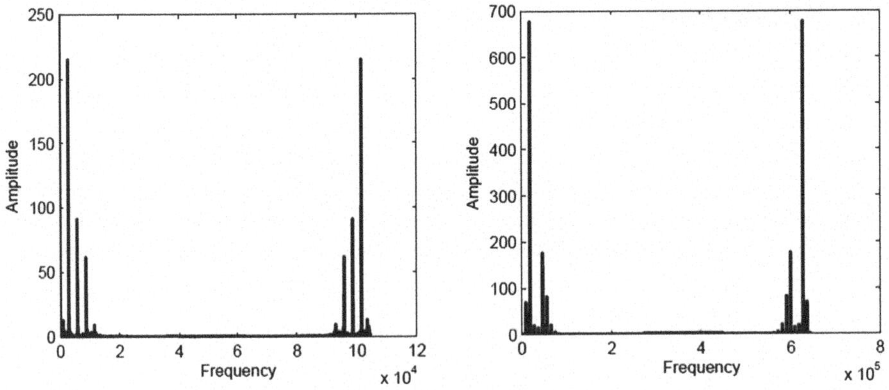

Fig. 9.10 FFT of single note played by a flute and a violin

Fig. 9.11 Three-dimensional space representation for PSO and MPSO

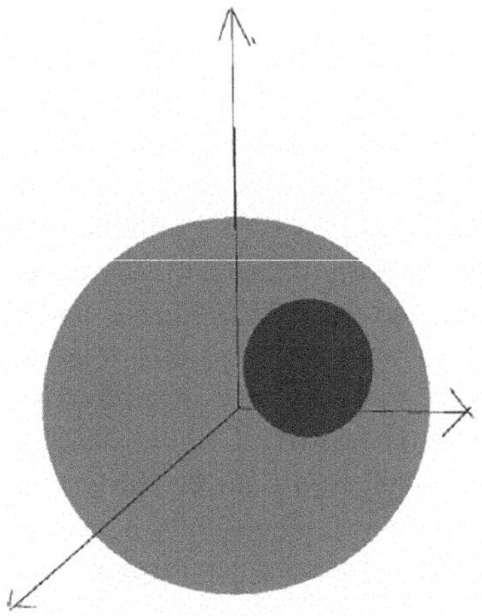

with minimum error, although not zero error. The filter coefficients corresponding to this point are used as the new initialization values for the second set of iterations. During the second step, the particles are assigned positions around the convergence point obtained from the first iteration. In Fig. 9.11, the lightly shaded area represents the space for ordinary PSO, while the darkly shaded area represents the space for modified PSO (MPSO). This iterative process can be repeated until the error approaches zero.

The order of the FIR filter for which the minimum error occurs is 15 for the flute and 13 for the violin. These filter coefficients generate the SSM wavelets for the flute and violin. The SSM wavelet for the flute is shown in Fig. 9.12. Figure 9.13 displays the original repeating waveform in a flute signal. The wavelet decompositionof a single note of the flute signal is performed using well-known wavelets and the SSM wavelet for the flute. In SSM wavelets, most of the signal energy is confined in the approximation coefficients rather than in the detail coefficients.

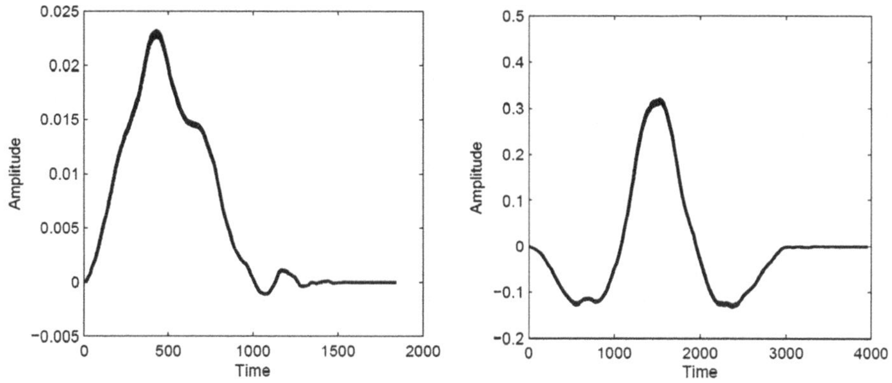

Fig. 9.12 SSM scaling and wavelet function for flute

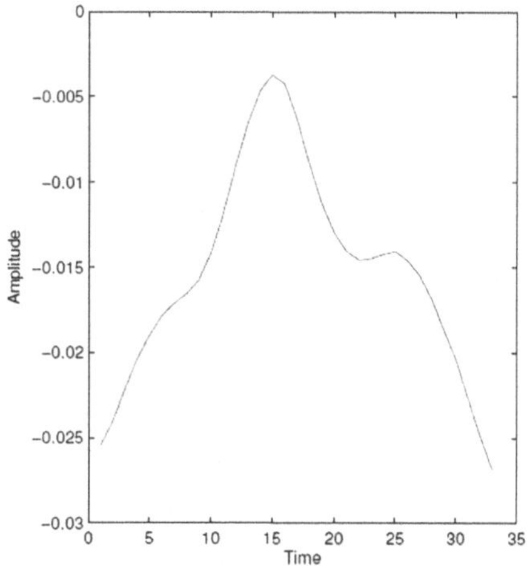

Fig. 9.13 Original repeating waveform in a flute signal

Design of Biorthogonal Wavelets

For designing biorthogonal wavelets, initially, the wavelet coefficients obtained using MPSO are normalized such that the sum of the squares of the coefficients becomes unity. As per the discussions in the previous sections, the number of coefficients in $h(n)$ and $\tilde{h}(n)$ should either be both even or both odd. Since the length of $h(n)$ for flute is 15, the length of $\tilde{h}(n)$ can be assumed to be 17. Thus,

$$h[n] = [0.2290, 0.1796, 0.0235, 0.2084, 0.2928, 0.1962, -0.0004, -0.0201, \\ 0.1703, 0.2593, -0.0139, -0.1915, 0.0359, 0.0728, -0.0244]$$

If the linear filter, $\tilde{h}(n)$, is linear phase, then

$$\tilde{h}(0) = \tilde{h}(16)$$
$$\tilde{h}(1) = \tilde{h}(15)$$
$$\tilde{h}(2) = \tilde{h}(14)$$
$$\tilde{h}(3) = \tilde{h}(13)$$
$$\tilde{h}(4) = \tilde{h}(12)$$
$$\tilde{h}(5) = \tilde{h}(11)$$
$$\tilde{h}(6) = \tilde{h}(10)$$
$$\tilde{h}(7) = \tilde{h}(9)$$
$$\tilde{h}(8) = \tilde{h}(8)$$

Since $\tilde{h}(n)$ is orthogonal to its shifts,

$$\sum_{k} \tilde{h}(n) = \sqrt{2}$$

i.e.,

$$2\tilde{h}(0)+2\tilde{h}(1)+2\tilde{h}(2)+2\tilde{h}(3)+2\tilde{h}(4)+2\tilde{h}(5)+2\tilde{h}(6)+2\tilde{h}(7)+\tilde{h}(8) = \sqrt{2} \quad (9.4)$$

For designing biorthogonal wavelets, initially, the wavelet coefficients obtained using MPSO are normalized such that the sum of the squares of the coefficients becomes unity. As per the discussions in the previous sections, the number of coefficients in $h(n)$ and $\tilde{h}(n)$ should either be both even or both odd. Since the length of $h(n)$ for flute is 15, the length of $\tilde{h}(n)$ can be assumed to be 17. Thus,

$$h[n] = [0.2290, 0.1796, 0.0235, 0.2084, 0.2928, 0.1962, -0.0004, -0.0201,$$
$$0.1703, 0.2593, -0.0139, -0.1915, 0.0359, 0.0728, -0.0244]$$

If the linear filter, $\tilde{h}(n)$, is linear phase, then

$$\tilde{h}(0) = \tilde{h}(16)$$
$$\tilde{h}(1) = \tilde{h}(15)$$
$$\tilde{h}(2) = \tilde{h}(14)$$
$$\tilde{h}(3) = \tilde{h}(13)$$
$$\tilde{h}(4) = \tilde{h}(12)$$
$$\tilde{h}(5) = \tilde{h}(11)$$
$$\tilde{h}(6) = \tilde{h}(10)$$
$$\tilde{h}(7) = \tilde{h}(9)$$
$$\tilde{h}(8) = \tilde{h}(8)$$

Since $\tilde{h}(n)$ is orthogonal to its shifts,

$$\sum_k \tilde{h}(n) = \sqrt{2}$$

i.e.,

$$2\tilde{h}(0)+2\tilde{h}(1)+2\tilde{h}(2)+2\tilde{h}(3)+2\tilde{h}(4)+2\tilde{h}(5)+2\tilde{h}(6)+2\tilde{h}(7)+\tilde{h}(8) = \sqrt{2} \quad (9.5)$$

Biorthogonality Conditions

Let $\tilde{h}(n)$ be nonzero in $N_1 \leq n \leq \tilde{N}_2$ and $h(n)$ be nonzero in $N_1 \leq n \leq N_2$. Assume $\tilde{h}(n)$ is nonzero in $0 \leq n \leq 16$. So, $\tilde{N}_1 = 0$ and $\tilde{N}_2 = 16$. N and $\tilde{N}$ are related by $N_2 - \tilde{N}_1 = 2k + 1$ and $\tilde{N}_2 - N_1 = 2\tilde{k} + 1$, where k and $\tilde{k}$ are integers. Substituting the values of $\tilde{N}_1$ and $\tilde{N}_2$ for $k = \tilde{k} = 7$, we get $N_1 = 1$ and $N_2 = 15$. So $h(n)$ is nonzero in $1 \leq n \leq 15$. The scaling function $h(n)$ is not orthogonal to its own translates. It should be orthogonal to the translates of $\tilde{h}(n)$. This is ensured by the condition

$$\sum_n h(n)\tilde{h}(n-2k) = \delta[k]$$

For $k = 0$,

$$\sum_n h(n)\tilde{h}(n) = 1$$

i.e.,

$$[h(0)+h(14)]\tilde{h}(1)+[h(1)+h(13)]\tilde{h}(2)+[h(2)+h(12)]\tilde{h}(3)+[h(3)+h(11)]\tilde{h}(4)$$
$$+[h(4)+h(10)]\tilde{h}(5)+[h(5)+h(9)]\tilde{h}(6)+[h(6)+h(8)]\tilde{h}(7)+h(7)\tilde{h}(8)=1$$

Substituting values

$$0.2046\tilde{h}(1)+0.2524\tilde{h}(2)+0.0594\tilde{h}(3)+0.0169\tilde{h}(4)+0.2789\tilde{h}(5)+0.4555\tilde{h}(6)$$
$$+0.1699\tilde{h}(7)-0.0201\tilde{h}(8)=1$$

$$0.2046\tilde{h}(1)+0.2524\tilde{h}(2)+0.0594\tilde{h}(3)+0.0169\tilde{h}(4)+0.2789\tilde{h}(5)$$
$$+0.4555\tilde{h}(6)+0.1699\tilde{h}(7)-0.0201\tilde{h}(8)=1 \tag{9.6}$$

For $k = 1$,

$$\sum_n h(n)\tilde{h}(n-2)=0$$

We get

$$h(1)\tilde{h}(0)+h(2)\tilde{h}(1)+h(3)\tilde{h}(2)+[h(4)+h(14)]\tilde{h}(3)+[h(5)+h(13)]\tilde{h}(4)$$
$$+[h(6)+h(12)]\tilde{h}(5)+[h(7)+h(11)]\tilde{h}(6)+[h(8)+h(10)]\tilde{h}(7)+h(9)\tilde{h}(8)=0$$

Substituting values

$$0.1796\tilde{h}(0)+0.0253\tilde{h}(1)+0.2084\tilde{h}(2)+0.2684\tilde{h}(3)+0.269\tilde{h}(4)$$
$$+0.0355\tilde{h}(5)-0.2116\tilde{h}(6)+0.1564\tilde{h}(7)+0.2593\tilde{h}(8)=0$$

For $k = 2$,

$$\sum_n h(n)\tilde{h}(n-4)=0$$

We get

$$h(3)\tilde{h}(0)+h(4)\tilde{h}(1)+h(5)\tilde{h}(2)+h(6)\tilde{h}(3)+h(7)\tilde{h}(4)+[h(8)+h(14)]\tilde{h}(5)$$
$$+[h(9)+h(13)]\tilde{h}(6)+[h(10)+h(12)]\tilde{h}(7)+h(11)\tilde{h}(8)=0$$

Substituting values

$$0.2084\tilde{h}(0)+0.2928\tilde{h}(1)+0.1962\tilde{h}(2)-0.0004\tilde{h}(3)-0.0201\tilde{h}(4)+0.1459\tilde{h}(5) + 0.3321\tilde{h}(6) + 0.022\tilde{h}(7) - 0.1915\tilde{h}(8) = 0$$

$$0.2084\tilde{h}(0) + 0.2928\tilde{h}(1) + 0.1962\tilde{h}(2) - 0.0004\tilde{h}(3) - 0.0201\tilde{h}(4) + 0.1459\tilde{h}(5) + 0.3321\tilde{h}(6) + 0.022\tilde{h}(7) - 0.1915\tilde{h}(8) = 0 \quad (9.7)$$

For $k = 3$,

$$\sum_n h(n)\tilde{h}(n-6) = 0$$

We get

$$h(5)\tilde{h}(0) + h(6)\tilde{h}(1) + h(7)\tilde{h}(2) + h(8)\tilde{h}(3) + h(9)\tilde{h}(4) + h(10)\tilde{h}(5) + h(11)\tilde{h}(6) + [h(12) + h(14)]\tilde{h}(7) + h(13)\tilde{h}(8) = 0$$

Substituting values

$$0.1962\tilde{h}(0)-0.0004\tilde{h}(1)-0.0201\tilde{h}(2)+0.1703\tilde{h}(3)+0.2593\tilde{h}(4)-0.0139\tilde{h}(5) - 0.1915\tilde{h}(6) + 0.0115\tilde{h}(7) + 0.0728\tilde{h}(8) = 0$$

$$0.1962\tilde{h}(0) - 0.0004\tilde{h}(1) - 0.0201\tilde{h}(2) + 0.1703\tilde{h}(3) + 0.2593\tilde{h}(4) - 0.0139\tilde{h}(5) - 0.1915\tilde{h}(6) + 0.0115\tilde{h}(7) + 0.0728\tilde{h}(8) = 0 \quad (9.8)$$

For $k = 4$,

$$\sum_n h(n)\tilde{h}(n-8) = 0$$

We get

$$h(7)\tilde{h}(0) + h(8)\tilde{h}(1) + h(9)\tilde{h}(2) + h(10)\tilde{h}(3) + h(11)\tilde{h}(4) + h(12)\tilde{h}(5) + h(13)\tilde{h}(6) + h(14)\tilde{h}(7) = 0$$

Substituting values

$$-0.0201\tilde{h}(0)+0.1703\tilde{h}(1)+0.2593\tilde{h}(2)-0.0139\tilde{h}(3)-0.1915\tilde{h}(4)+0.0359\tilde{h}(5) + 0.0728\tilde{h}(6) - 0.0244\tilde{h}(7) = 0$$

$$-0.0201\tilde{h}(0) + 0.1703\tilde{h}(1) + 0.2593\tilde{h}(2) - 0.0139\tilde{h}(3) - 0.1915\tilde{h}(4) + 0.0359\tilde{h}(5) + 0.0728\tilde{h}(6) - 0.0244\tilde{h}(7) = 0 \quad (9.9)$$

For $k = 5$,

$$\sum_n h(n)\tilde{h}(n-10) = 0$$

We get

$$h(9)\tilde{h}(0) + h(10)\tilde{h}(1) + h(11)\tilde{h}(2) + h(12)\tilde{h}(3) + h(13)\tilde{h}(4) + h(14)\tilde{h}(5) = 0$$

Substituting values

$$-0.1915\tilde{h}(0)-0.0139\tilde{h}(1)+0.2593\tilde{h}(2)+0.1703\tilde{h}(3)-0.0201\tilde{h}(4)+0.2286\tilde{h}(5) + 0.3758\tilde{h}(6) + 0.3163\tilde{h}(7) + 0.2084\tilde{h}(8) = 0$$

$$-0.1915\tilde{h}(0) - 0.0139\tilde{h}(1) + 0.2593\tilde{h}(2) + 0.1703\tilde{h}(3) - 0.0201\tilde{h}(4) + 0.2286\tilde{h}(5) + 0.3758\tilde{h}(6) + 0.3163\tilde{h}(7) + 0.2084\tilde{h}(8) = 0 \quad (9.10)$$

For $k = -3$,

$$\sum_n h(n)\tilde{h}(n+6) = 0$$

We get

$$h(9)\tilde{h}(0) + h(8)\tilde{h}(1) + h(7)\tilde{h}(2) + h(6)\tilde{h}(3) + h(5)\tilde{h}(4) + h(4)\tilde{h}(5) + h(3)\tilde{h}(6) + [h(2) + h(0)]\tilde{h}(7) + h(1)\tilde{h}(8) = 0$$

Substituting values

$$0.2593\tilde{h}(0)+0.1703\tilde{h}(1)-0.0201\tilde{h}(2)-0.0004\tilde{h}(3)+0.1962\tilde{h}(4)-0.2928\tilde{h}(5) + 0.2084\tilde{h}(6) + 0.2525\tilde{h}(7) + 0.1796\tilde{h}(8) = 0$$

$$0.2593\tilde{h}(0)+0.1703\tilde{h}(1)-0.0201\tilde{h}(2)-0.0004\tilde{h}(3)+0.1962\tilde{h}(4)-0.2928\tilde{h}(5) + 0.2084\tilde{h}(6) + 0.2525\tilde{h}(7) + 0.1796\tilde{h}(8) = 0 \quad (9.11)$$

For $k = -4$,

$$\sum_n h(n)\tilde{h}(n+8) = 0$$

Hence,

$$h(7)\tilde{h}(0) + h(6)\tilde{h}(1) + h(5)\tilde{h}(2) + h(4)\tilde{h}(3) + h(3)\tilde{h}(4) + h(2)\tilde{h}(5) + h(1)\tilde{h}(6) + h(0)\tilde{h}(7) = 0$$

Substituting values

$$-0.0201\tilde{h}(0)-0.0004\tilde{h}(1)+0.1962\tilde{h}(2)+0.2928\tilde{h}(3)+0.2084\tilde{h}(4)+0.0235\tilde{h}(5) + 0.1796\tilde{h}(6) + 0.2290\tilde{h}(7) = 0$$

$$-0.0201\tilde{h}(0)-0.0004\tilde{h}(1)+0.1962\tilde{h}(2)+0.2928\tilde{h}(3)+0.2084\tilde{h}(4) +0.0235\tilde{h}(5) + 0.1796\tilde{h}(6) + 0.2290\tilde{h}(7) = 0 \quad (9.12)$$

For $k = -5$,

$$\sum_n h(n)\tilde{h}(n+10) = 0$$

Thus,

$$h(5)\tilde{h}(0) + h(4)\tilde{h}(1) + h(3)\tilde{h}(2) + h(2)\tilde{h}(3) + h(1)\tilde{h}(4) + h(0)\tilde{h}(5) = 0$$

Substituting values

$$0.1962\tilde{h}(0) + 0.2928\tilde{h}(1) + 0.2084\tilde{h}(2) + 0.0235\tilde{h}(3) + 0.1796\tilde{h}(4) + 0.2290\tilde{h}(5) = 0$$

For $k = -5$:

$$\sum_n h(n) \cdot h(n+10) = 0 \tag{9.13}$$

Thus,

$$h(5) \cdot h(0) + h(4) \cdot h(1) + h(3) \cdot h(2) + h(2) \cdot h(3) + h(1) \cdot h(4) + h(0) \cdot h(5) = 0 \tag{9.14}$$

Substituting values,

$$0.1962 \cdot h(0) + 0.2928 \cdot h(1) + 0.2084 \cdot h(2) + 0.0235 \cdot h(3) + 0.1796 \cdot h(4) + 0.2290 \cdot h(5) = 0 \tag{9.15}$$

For $k = -6$:

$$\sum_n h(n) \cdot h(n+12) = 0 \tag{9.16}$$

Therefore,

$$h(3) \cdot h(0) + h(2) \cdot h(1) + h(1) \cdot h(2) + h(0) \cdot h(3) = 0 \tag{9.17}$$

Substituting,

$$0.2084 \cdot h(0) + 0.0235 \cdot h(1) + 0.1796 \cdot h(2) + 0.2290 \cdot h(3) = 0 \tag{9.18}$$

For $k = -7$,

$$\sum_n h(n)\tilde{h}(n+14) = 0$$

Thus, we get

$$h(1)\tilde{h}(0) + h(0)\tilde{h}(1) = 0$$

Substituting values,

$$0.1796\tilde{h}(0) - 0.2290\tilde{h}(1) = 0$$

$$0.1796\tilde{h}(0) - 0.2290\tilde{h}(1) = 0 \tag{9.19}$$

Applying the vanishing moment condition,

$$\int Z(t)\,dt = 0 \Longrightarrow \sum_k (-1)^n \tilde{h}(N-k-1) = 0$$

So we get

$$2\tilde{h}(0) - 2\tilde{h}(1) + 2\tilde{h}(2) - 2\tilde{h}(3) + 2\tilde{h}(4) - 2\tilde{h}(5) + 2\tilde{h}(6) - 2\tilde{h}(7) + \tilde{h}(8) = 0$$

$$2\tilde{h}(0)-2\tilde{h}(1)+2\tilde{h}(2)-2\tilde{h}(3)+2\tilde{h}(4)-2\tilde{h}(5)+2\tilde{h}(6)-2\tilde{h}(7)+\tilde{h}(8) = 0 \quad (9.20)$$

Similarly,

$$-0.0201\tilde{h}(0)+0.3749\tilde{h}(1)+0.5117\tilde{h}(2)+0.0455\tilde{h}(3)-0.1746\tilde{h}(4)+0.3148\tilde{h}(5) + 0.5283\tilde{h}(6) + 0.1455\tilde{h}(7) - 0.0201\tilde{h}(8) = 1$$

$$\begin{aligned} &- 0.0201\tilde{h}(0) + 0.3749\tilde{h}(1) + 0.5117\tilde{h}(2) + 0.0455\tilde{h}(3) - 0.1746\tilde{h}(4) \\ &+ 0.3148\tilde{h}(5) + 0.5283\tilde{h}(6) + 0.1455\tilde{h}(7) - 0.0201\tilde{h}(8) = 1 \end{aligned} \quad (9.21)$$

Now, we have nine equations in nine unknowns. These set of linear equations can be expressed in matrix form as $AX = Y$ and can be solved using $X = A^{-1}Y$, where A is the coefficient matrix given by... The system of equations can be expressed in matrix form as follows:

$$A = \begin{bmatrix} 2 & 2 & 2 & 2 & 2 & 2 & 2 & 2 & 1 \\ 0.1796 & 0.0235 & 0.2084 & 0.2684 & 0.269 & 0.0355 & -0.2116 & 0.1564 & 0.2593 \\ 0.2084 & 0.2928 & 0.1962 & -0.0004 & -0.0201 & 0.1459 & 0.3321 & 0.022 & -0.1915 \\ 0.1962 & -0.0004 & -0.0201 & 0.1703 & 0.2593 & -0.0139 & -0.1915 & 0.0115 & 0.0728 \\ 0.0728 & 0.0359 & -0.1915 & 0.2151 & 0.4389 & 0.1938 & 0.1883 & 0.2924 & 0.1962 \\ -0.1915 & -0.0139 & 0.2593 & 0.1703 & -0.0201 & 0.2286 & 0.3758 & 0.3163 & 0.2084 \\ 0.2593 & 0.1703 & -0.0201 & -0.0004 & 0.1962 & 0.2928 & 0.2084 & 0.2525 & 0.1796 \\ 2 & -2 & 2 & -2 & 2 & -2 & 2 & -2 & 1 \\ -0.0201 & 0.3749 & 0.5117 & 0.0455 & -0.1746 & 0.3148 & 0.5283 & 0.1455 & -0.0201 \\ 1 & & & & & & & & \end{bmatrix}$$

The matrix Y is given by

$$Y = \begin{bmatrix} 0 \\ 0 \\ 0 \\ 0 \\ 0 \\ 0 \\ 0 \\ 0 \\ 1 \\ 1 \end{bmatrix}$$

By solving the system, we get X as

$$X = \begin{bmatrix} -2.1465 \\ 2.9536 \\ -0.5659 \\ 1.4113 \\ 0.4386 \\ 2.9206 \\ 0.6092 \\ -6.9320 \\ 4.0364 \\ 1 \end{bmatrix}$$

Thus, the coefficients $\tilde{h}(n)$ are given by... The coefficients $\tilde{h}(n)$ are obtained as follows:

$$\tilde{h}(n) = \begin{bmatrix} -2.1465 & 2.9536 & -0.5659 & 1.4113 & 0.4386 & 2.9206 & 0.6092 & -6.9320 & 4.0364 \\ -6.9320 & 0.6092 & 2.9206 & 0.4386 & 1.4113 & -0.5659 & 2.9536 & -2.1465 & \end{bmatrix}$$

The coefficients $g(n)$ are obtained from $\tilde{h}(n)$ by

$$g(n) = \sum_k (-1)^n \tilde{h}(N - k - 1)$$

Hence, $g(n)$ is given by

$$g(n) = \begin{bmatrix} -2.1465 & -2.9536 & -0.5659 & -1.4113 & 0.4386 & -2.9206 & 0.6092 & 6.9320 & 4.0364 \\ 6.9320 & 0.6092 & -2.9206 & 0.4386 & -1.4113 & -0.5659 & -2.9536 & -2.1465 & \end{bmatrix}$$

Similarly, the coefficients $\tilde{g}(n)$ are obtained from $h(n)$ by

$$\tilde{g}(n) = \sum_{k} (-1)^n h(N - k - 1)$$

Thus, $\tilde{g}(n)$ is given by

$$\tilde{g}(n) = \begin{bmatrix} -0.0244 \ -0.0728 \ \ 0.0359 \ \ 0.1915 \ -0.0139 \ -0.2593 \ 0.1703 \ 0.0201 \ -0.0004 \\ -0.1962 \ \ 0.2928 \ \ -0.2084 \ 0.0235 \ -0.1796 \ \ 0.2290 \end{bmatrix}$$

9.7 Solved Problems

Problem Consider a noisy audio signal sampled at $f_s = 44100$ Hz. Design a low-pass filter using a genetic algorithm (GA) to minimize the mean squared error (MSE) between the filtered signal and a reference clean signal. Optimize the filter's cutoff frequency and order, given the bounds:

- Cutoff frequency: [100 Hz, 8000 Hz]
- Filter order: [1, 10]

Write a MATLAB function to implement this, and compute the optimized parameters and the MSE.

Solution:

```
% Load or generate the noisy and clean signals

    % Generate a clean signal and add noise

    % Initialize GA Parameters


    % Initialize Population
    population = initializePopulation(populationSize, bounds);

    % Evaluate Initial Fitness
    fitness = zeros(populationSize, 1);
    for i = 1:populationSize
    fitness(i) = denoiseObjective(noisySignal, cleanSignal, fs, population(i, :)
        );
    end

    % Main GA Loop
    for gen = 1:numGenerations
    newPopulation = zeros(size(population));

    for i = 1:populationSize
    % Selection (Tournament)
    parent1 = tournamentSelection(population, fitness, 3);
```

```
parent2 = tournamentSelection(population, fitness, 3);

% Crossover
if rand < crossoverRate
child = singlePointCrossover(parent1, parent2);
else
child = parent1;
end

% Mutation
if rand < mutationRate
child = mutate(child, bounds);
end

% Evaluate Child Fitness
newPopulation(i, :) = child;
end

% Evaluate Fitness of New Population
for i = 1:populationSize
fitness(i) = denoiseObjective(noisySignal, cleanSignal, fs, newPopulation(i,
    :));
end

% Update Population
population = newPopulation;

% Display Progress
[bestFitness, bestIdx] = min(fitness);
disp(['Generation ', num2str(gen), ': Best Fitness = ', num2str(bestFitness)
    ]);
end

% Final Optimized Parameters
[bestFitness, bestIdx] = min(fitness);
optimizedParams = population(bestIdx, :);
disp(['Optimized Parameters: Cutoff Frequency = ', num2str(optimizedParams
    (1)), ...
', Order = ', num2str(optimizedParams(2))]);

% Apply Filter with Optimized Parameters
filteredSignal = applyFilter(noisySignal, fs, optimizedParams);

% Plot Results


% Objective Function
function mse = denoiseObjective(noisySignal, cleanSignal, fs, params)
cutoffFreq = params(1);
order = round(params(2));
filteredSignal = applyFilter(noisySignal, fs, [cutoffFreq, order]);
mse = mean((filteredSignal - cleanSignal).^2);
end
```

```
% Filter Application
function filteredSignal = applyFilter(noisySignal, fs, params)
cutoffFreq = params(1);
order = round(params(2));
[b, a] = butter(order, cutoffFreq / (fs / 2), 'low');
filteredSignal = filter(b, a, noisySignal);
end

% Initialize Population
function population = initializePopulation(populationSize, bounds)
population = bounds(1, :) + rand(populationSize, size(bounds, 2)) .* (bounds(2, :) - bounds(1, :));
end

% Tournament Selection
function selected = tournamentSelection(population, fitness, tournamentSize)
competitors = randi(size(population, 1), tournamentSize, 1);
[~, bestIdx] = min(fitness(competitors));
selected = population(competitors(bestIdx), :);
end

% Single-Point Crossover
function child = singlePointCrossover(parent1, parent2)
point = randi(length(parent1));
child = [parent1(1:point), parent2(point+1:end)];
end

% Mutation
function mutated = mutate(individual, bounds)
geneIdx = randi(length(individual));
mutationValue = (bounds(2, geneIdx) - bounds(1, geneIdx)) * randn * 0.1;
individual(geneIdx) = individual(geneIdx) + mutationValue;
individual(geneIdx) = max(bounds(1, geneIdx), min(bounds(2, geneIdx), individual(geneIdx))); % Ensure bounds
mutated = individual;
end
```

Problem Using the same noisy and clean audio signals as in Exercise 1, design a low-pass filter using differential evolution (DE). The optimization should minimize the mean squared error (MSE) between the denoised and clean signals. Write a MATLAB function to implement this, and compute the optimized parameters.

Solution:

```
clc; clear; close all;

% Load or generate the noisy and clean signals
fs = 44100; % Sampling frequency
duration = 5; % Duration in seconds
t = (0:1/fs:duration-1/fs)'; % Time vector

% Generate a clean signal and add noise
```

```
cleanSignal = sin(2 * pi * 440 * t); % Example clean signal (A4 note)
noisySignal = cleanSignal + 0.1 * randn(size(cleanSignal)); % Add white
    noise

% DE Parameters
populationSize = 50;
numGenerations = 50;
mutationFactor = 0.8;
crossoverRate = 0.9;
bounds = [100, 1; 8000, 10]; % [Cutoff frequency; Order]

% Initialize Population
population = initializePopulation(populationSize, bounds);

% Main DE Loop
for gen = 1:numGenerations
newPopulation = zeros(size(population));

for i = 1:populationSize
% Mutation and Crossover
idxs = randperm(populationSize, 3); % Random selection of 3 individuals
base = population(idxs(1), :);
mutant = population(idxs(2), :) + mutationFactor * (population(idxs(3), :) -
    population(idxs(2), :));
trial = crossover(base, mutant, crossoverRate, bounds);

% Evaluate Fitness
trialFitness = evaluateFitness(noisySignal, cleanSignal, fs, trial);
baseFitness = evaluateFitness(noisySignal, cleanSignal, fs, base);

% Selection
if trialFitness < baseFitness
newPopulation(i, :) = trial;
else
newPopulation(i, :) = base;
end
end

% Update Population
population = newPopulation;

% Display Progress
[bestFitness, bestIdx] = min(arrayfun(@(i) evaluateFitness(noisySignal,
    cleanSignal, fs, population(i, :)), 1:populationSize));
disp(['Generation ', num2str(gen), ': Best Fitness = ', num2str(bestFitness)
    ]);
end

% Final Optimized Parameters
[bestFitness, bestIdx] = min(arrayfun(@(i) evaluateFitness(noisySignal,
    cleanSignal, fs, population(i, :)), 1:populationSize));
optimizedParams = population(bestIdx, :);
```

```
disp(['Optimized Parameters: Cutoff Frequency = ', num2str(optimizedParams
    (1)), ...
', Order = ', num2str(optimizedParams(2))]);

% Apply Filter with Optimized Parameters
filteredSignal = applyFilter(noisySignal, fs, optimizedParams);

% Plot Results
figure;
subplot(3, 1, 1); plot(t, cleanSignal); title('Clean Signal');
subplot(3, 1, 2); plot(t, noisySignal); title('Noisy Signal');
subplot(3, 1, 3); plot(t, filteredSignal); title('Filtered Signal');

% Supporting Functions

% Evaluate Fitness (Objective Function)
function mse = evaluateFitness(noisySignal, cleanSignal, fs, params)
filteredSignal = applyFilter(noisySignal, fs, params);
mse = mean((filteredSignal - cleanSignal).^2);
end

% Apply Filter
function filteredSignal = applyFilter(noisySignal, fs, params)
cutoffFreq = params(1);
order = round(params(2));
[b, a] = butter(order, cutoffFreq / (fs / 2), 'low');
filteredSignal = filter(b, a, noisySignal);
end

% Initialize Population
function population = initializePopulation(populationSize, bounds)
population = bounds(1, :) + rand(populationSize, size(bounds, 2)) .* (bounds
    (2, :) - bounds(1, :));
end

% Crossover Operator
function trial = crossover(base, mutant, crossoverRate, bounds)
trial = base;
for i = 1:length(base)
if rand < crossoverRate
trial(i) = mutant(i);
end
trial(i) = max(bounds(1, i), min(bounds(2, i), trial(i))); % Ensure bounds
end
end
```

Problem Using the same noisy and clean audio signals as in Exercise 1, design a low-pass filter using particle swarm optimization (PSO). The optimization should minimize the mean squared error (MSE) between the denoised and clean signals. Write a MATLAB function to implement this, and compute the optimized parameters and the MSE.

Solution:

```
% Generate a clean signal and add noise

% Initialize PSO Parameters



% Initialize Population
population = initializePopulation(numParticles, bounds);
velocity = zeros(size(population));

% Main PSO Loop
for gen = 1:numGenerations
for i = 1:numParticles
% Evaluate Fitness
fitness(i) = evaluateFitness(noisySignal, cleanSignal, fs, population(i, :))
    ;

% Update Personal Best
if fitness(i) < personalBestFitness(i) || gen == 1
personalBest(i, :) = population(i, :);
personalBestFitness(i) = fitness(i);
end

% Update Global Best
[globalBestFitness, bestIdx] = min(personalBestFitness);
globalBest = personalBest(bestIdx, :);

% Update Velocity
velocity(i, :) = w * velocity(i, :) + ...
c1 * rand() * (personalBest(i, :) - population(i, :)) + ...
c2 * rand() * (globalBest - population(i, :));

% Update Position
population(i, :) = population(i, :) + velocity(i, :);

% Ensure the population stays within bounds
population(i, :) = max(bounds(1, :), min(bounds(2, :), population(i, :)));
end

% Display Progress
disp(['Generation ', num2str(gen), ': Best Fitness = ', num2str(
    globalBestFitness)]);
end

% Final Optimized Parameters
disp(['Optimized Parameters: Cutoff Frequency = ', num2str(globalBest(1)),
    ...
```

```
', Order = ', num2str(globalBest(2))]);

% Apply Filter with Optimized Parameters
filteredSignal = applyFilter(noisySignal, fs, globalBest);

% Plot Results



% Evaluate Fitness (Objective Function)
function mse = evaluateFitness(noisySignal, cleanSignal, fs, params)
filteredSignal = applyFilter(noisySignal, fs, params);
mse = mean((filteredSignal - cleanSignal).^2);
end

% Apply Filter
function filteredSignal = applyFilter(noisySignal, fs, params)
cutoffFreq = params(1);
order = round(params(2));
[b, a] = butter(order, cutoffFreq / (fs / 2), 'low');
filteredSignal = filter(b, a, noisySignal);
end

% Initialize Population
function population = initializePopulation(numParticles, bounds)
population = bounds(1, :) + rand(numParticles, size(bounds, 2)) .* (bounds(2, :) - bounds(1, :));
end
```

Bibliography

Chapa, J. O., & Rao, R. M. (2000). Algorithms for designing wavelets to match a specified signal. *IEEE Transactions on Signal Processing, 48*(12), 3395–3406. https://doi.org/10.1109/78.887001.

Daubechies, I. (1992). Ten lectures on wavelets. Society for Industrial and Applied Mathematics.

Daubechies, I., & Sweldens, W. (1996). Factoring wavelet transforms into lifting steps. Technical report, Princeton and Lucent Technologies, NJ. Preprint.

Ghanbari, M. (2003). *Standard codecs: Image compression to advanced video coding*. Institution of Electrical Engineers.

Luo, G., & Zhang, D. (2012). Wavelet denoising. In *Advances in wavelet theory and their applications in engineering, physics and technology*. InTech. https://doi.org/10.5772/37424.

Mallat, S. (2008). *A wavelet tour of signal processing: The sparse way* (3rd ed.). Academic Press.

Sharma, A., Achuth, P. V., Pachori, R. B., & Gadre, V. M. (2017). A parametrization technique to design joint time–frequency optimized discrete-time biorthogonal wavelet bases. *Signal Processing, 135*, 107–120. https://doi.org/10.1016/j.sigpro.2016.12.019.

Shim, K., Garofalakis, M., & Rastogi, R. (2001). Approximate query processing using wavelets. *The VLDB Journal, 10(2–3)*, 199–223.

Sinith, M. S. (2017). Modelling and analysis of Indian Classical Music for raga recognition, Ph.D. dissertation. Department of Electronics and Communication Engineering, Amrita Vishwa Vidyapeetham, Amrita School of Engineering, Bangalore, India.

Soman, K. P., & Ramachandran, K. I. (2005). *Insight into wavelets: From theory to practice* (2nd ed.). PHI.

Strang, G., & Nguyen, T. (1997). *Wavelets and filter banks*. Wellesley-Cambridge Press.

Taubman, D. S., & Marcellin, M. W. (2002). *JPEG2000: Image compression fundamentals, standards and practice*. Springer.

Vaidyanathan, P. P. (1992). *Multirate systems and filter banks*. Prentice Hall.

Zhang, Z., Lu, Y., & Hsu, H. (2007). Detecting surface geostrophic currents using wavelet filter from satellite geodesy. *Science China Earth Sciences, 50*, 918–926.

M. S. Sinith et al., *Wavelets and Multi-Resolution Analysis*,
https://doi.org/10.1007/978-3-031-92750-8